The Passivhaus Handbook

The Passivhaus Handbook

A practical guide to constructing and retrofitting buildings for ultra-low energy performance

Janet Cotterell and Adam Dadeby

green books

First published in 2012 by Green Books

Reprinted in 2013 by UIT / Green Books
PO Box 145, Cambridge CB4 1GQ, UK
www.greenbooks.co.uk

© 2012 Janet Cotterell and Adam Dadeby
All rights reserved
The authors have asserted their moral rights under the Copyright, Designs and Patents Act 1988

No part of this book may be used or reproduced in any manner without written permission, except in the case of brief quotations in critical articles or reviews.

Design by Jayne Jones

Photographs not otherwise credited are by the authors. Many of these were taken at the Totnes Passivhaus.

ISBN: 978 0 85784 019 6 (paperback)
ISBN: 978 0-85784 116 2 (ePub)
ISBN: 978 0 85784 115 5 (pdf)

10 9 8 7 6 5 4 3 2

Cover image: Totnes Passivhaus. *Image: Malcolm Baldwin*. Back cover images: Left: Camden Passivhaus. *Image: Jefferson Smith*; Centre: SurPlus-Home. *Image: ENERGATE®*; Right: Underhill. *Image: Samuel Ashfield*

Disclaimer: At the time of going to press, the advice and information in this book are believed to be true and accurate. The author and publishers accept no liability for actions inspired by this book.

Architectural credits for UK buildings pictured in this book are detailed in Appendix D, pages 234-8. Those for non-UK buildings are as follows:
Page 10: Bilyeu Homes, Inc. **Page 26:** Corey Saft. **Page 28:** Top right: Team Germany / Technische Universität Darmstadt. Top left: Aslett Naftel and consultants BRE. Middle left: Madreiter Architects. Bottom: Simmonds.Mills. **Page 82:** Passivhaus Homes Ltd / CTT Sustainable Architect. **Page 194:** Garnet Vallentin.

Details of uncaptioned images throughout the book:
Page 2: Denby Dale, Yorkshire: Passivhaus using standard cavity-wall construction with stone outer leaf. *Image: Green Building Store*
This page: Crossway, Kent: Passivhaus using local materials and a parabolic vault roof, finalist in 2009 *Grand Designs* Award. *Image: Richard Hawkes*

Page 10: Rue-Evans house, Oregon, USA: Traditional house using natural materials; first Passivhaus on west coast USA. *Image: Sarah Evans*
Page 14: Canolfan Hyddgen, Machynlleth, Wales: Training and education centre for Powys County Council, BREEAM 'Excellent' standard. First Passivhaus building in the UK. *Image: JPW Construction*

Pages 88-9: Tigh-na-Cladach, Dunoon, Scotland: Affordable housing and first Passivhaus in Scotland. *Image: Andrew Lee*
Page 211: *Image: Malcolm Baldwin*
Page 213: *Image: Morgan O'Driscoll Photography*
Page 214: Grove Cottage: *Image: Simmonds.Mills*; KlimaSolarHaus: *Image: Henrietta Lynch*

Contents

Foreword by Wolfgang Feist 9
Introduction 11
How to use this book 13

Part One: The how and why of Passivhaus

Chapter 1   What is a Passivhaus? 16
The Passivhaus energy standard, the Passivhaus concept, common misconceptions, retrofitting and the EnerPHit energy standard

Chapter 2   The economics of a Passivhaus 32
Economic culture, energy costs and supplies, factors affecting property prices, factors affecting Passivhaus costs, methods of determining cost-effectiveness

Chapter 3   Passivhaus Certification 40
Certification of buildings, building components, designers and contractors; the process and information required for building certification

Chapter 4   Challenges of meeting the Passivhaus standard 46
UK building culture, education and employment in construction, professional expertise, risk-aversity, the planning system, vernacular styles, the team approach, the disadvantages of a Passivhaus, Passivhaus and other low-energy standards

Chapter 5   Natural materials, zero carbon and resilience 60
Natural and low-embodied-energy building materials, zero carbon and the Code for Sustainable Homes (CSH), on-site low- or zero-carbon energy, post-peak energy, energy returned on energy invested (EROEI)

Chapter 6   Setting up a Passivhaus project 74
Choosing a plot, planning considerations, retrofit considerations, phased retrofitting and extensions, selecting an architect and builder, the role of the client

Part Two: Passivhaus projects: a practical guide

Chapter 7   Using the Passivhaus Planning Package (PHPP) 90
History of the PHPP, PHPP worksheets: Verification, U-Values, Ground, WinType, Windows, Shading, Ventilation, Annual Heating Demand, Summer, Shading-S, DHW + Distribution, SolarDHW, Climate

Chapter 8   Thermal bridges 110
Constructional and geometrical thermal bridges, linear and point thermal bridges, thermal bypass, internal and external psi-values, dealing with thermal bridges, thermal bridge calculation

| Chapter 9 | Airtightness and sequencing | 122 |

Air leakage and Passivhaus, internal air quality (IAQ), airtightness standards, wind-tightness, breathable materials, airtight materials, air leakage at the design stage and construction stage, sequencing, on-site communication and training, airtightness testing, typical airtight construction details

| Chapter 10 | Moisture | 144 |

Liquid moisture and water vapour, relative humidity (RH) and indoor air quality (IAQ), capillarity, hygroscopicity, vapour permeability, moisture management in construction, breathability, example constructions (new build and retrofit)

| Chapter 11 | Windows | 168 |

High-performance windows, U-values (frame, pane, spacer and installation factors), solar gain, window installation, window sills, doors, roof lights, avoiding summer overheating, the construction phase, future developments and costs

| Chapter 12 | Ventilation | 188 |

Ventilation in UK housing and in Passivhaus, indoor air quality (IAQ), humidity, mechanical ventilation with heat recovery (MVHR) (components, heating, efficiency, noise levels, possible objections to MVHR, installation skills)

| Chapter 13 | Living in a Passivhaus | 208 |

Noise, energy bills, kitchen and bathrooom, drying clothes, the MVHR, entering and leaving the house, case studies (Totnes Passivhaus, Denby Dale, Grove Cottage, Passivhaus apartment buildings)

| Chapter 14 | Policy change in the UK | 216 |

Planning, a building-fabric-based energy standard, floor measurement conventions, VAT, Energy Performance Certificates, property tax, change in the construction sector, self-build, home-grown Passivhaus products, culture and policy-making

Appendix A: Space heating and hot water 230
Appendix B: Thermal conductivity values 232
Appendix C: US units – metric conversions 233
Appendix D: Certified Passivhaus projects in the UK 234

Glossary of terms 239
Glossary of units 245

Notes 246
Resources 249
Index 252

This book is dedicated to Erica Aslett
and Christine Ann Harrison

Acknowledgements

We would like to thank the various people who helpfully commented on specific chapters, as well as those who gave their time, advice and encouragement. In particular:

Moisture	**Niall Crosson** of Ecological Building Systems Ltd
	Anna Carton for background research and general input
Ventilation	**David McHugh** of Pro-Air Systems
Standard Assessment Procedure (SAP) and Code for Sustainable Homes (CSH)	**Phil Neve** of AABEN
Certification and thermal bridges	**Peter Warm** of WARM: Low Energy Building Practice
Windows	**Mathias Häußler** of ENERGATE® windows; **Conor Ryan** of Ambiwood
Planning	**Steve Munday, Roger Estop**
Living in a Passivhaus	**Henrietta Lynch**
Front cover image	**Malcolm Baldwin**
General input, support and patience!	**Jonathan Williams** of Williams and Partners
Editing the book and managing us!	**Amanda Cuthbert** and all the helpful team at Green Books, particularly **Alethea Doran** for her attention to detail

Foreword

A Passivhaus manual for the English-speaking world has been long overdue. I am impressed by the swift uptake of Passivhaus in the UK and North America and am delighted that, in this handbook, architects, practitioners and all those wishing to build a Passivhaus will be able to find what they need to successfully carry out their projects.

As the scientific head of the Passivhaus Institut, I should perhaps highlight one point: *Passivhaus is open source*. Twenty-four years ago, Bo Adamson and I developed the concept and built the first prototype. This was part of a study to see just how far we could take low-energy concepts. The results were overwhelmingly positive: the building performed almost exactly as planned and, despite initial scepticism, the residents of the first terraced houses were not only satisfied but truly excited by the outcome.

We have published all principles, tools, details and design aids used, as well as all monitoring results, and worked hard to ensure the standard is accessible to all – the PHPP calculation tool, for example, is 100-per-cent transparent. Anyone may choose to build a Passivhaus, without asking permission (beyond compliance with local building codes). That said, it is important not to take the task of building a Passivhaus lightly: given that the criteria are published, the necessary planning tools are widely available and there is a diversity of functional, built examples, it goes without saying that a building really should meet the Passivhaus standard if labelled as such. Our voluntary certification scheme is there to provide this important quality assurance. This book will complement the support provided by procedures such as building certification (a step that can result in savings far beyond what the certification itself costs) and by tools such as the PHPP.

The availability of Passivhaus criteria for individual building components, especially for use in retrofits, makes it clear to the building sector which goals can be economically and sensibly achieved. Passivhaus is part of a critical paradigm shift, whereby the value once provided by oil, coal and natural gas in our buildings is instead provided by long-lasting, quality products such as windows, ventilation systems and insulation. This new form of value may be provided locally, thus offering opportunities to small businesses: Passivhaus is not only a positive contribution to the energy revolution but also to the formation of a sustainable and stable economy.

In the UK, there has been much discussion about whether a shift to a sustainable energy supply without the need for coal and uranium is even realistic. Passivhaus shows that this shift is feasible and also offers an array of further benefits, not least higher-quality buildings and improved comfort.

It is my hope that this handbook finds many readers. In turn, I wish its readers every success in carrying out their Passivhaus projects.

Wolfgang Feist
Scientific Director, Passivhaus Institut
October 2012

Introduction

This book is aimed at a broad readership, including builders, self-builders, architects, energy assessors, registered social landlords (RSLs), planners, building control officers, politicians, clients – in fact anyone who is, wants to be or should be involved in future housing provision.

It is intended to provide knowledge of both the methodology and the skills needed to achieve genuinely low-energy buildings, whether new or retrofitted, that perform as intended.[1] In the UK especially we have a very poor record when it comes to transferring design intent into real-world building performance. Building sustainably involves a wide variety of considerations, from water use to recycling to low-embodied-energy materials, and so forth. Passivhaus focuses on the energy consumed in the lifetime of a building. This is where we make (by far) the largest carbon savings – a necessary step if we seriously want to move towards a low-carbon economy and address a less energy-rich future. This is also where we have failed to find success to date in our houses and eco-buildings. We can install solar panels on our roofs, we can reduce water use, we can recycle, we can choose to use materials with lower embodied energy. But can we radically reduce our space-heating needs?

Effectively reducing the energy consumed for space heating (or, in hotter climates, cooling) requires training and up-skilling, both of which pose significant challenges. In the UK, we are in the process of adopting a broad-based sustainability assessment system, the Code for Sustainable Homes (CSH), where points are awarded for a range of possible sustainable features – you collect the points and rise up the levels (from Levels 3 to 6, where Levels 5 and 6 are currently termed 'zero carbon'). While the broad approach has much merit, an unintended side effect can be that fabric performance (the energy demands of the building itself) is perceived as one option among many – and other criteria are often easier, quicker and less challenging (and more exciting!) to meet. Energy in use, however, lies at the very heart of a truly sustainable architecture. If you get this right it also brings many associated benefits – health (indoor air quality), material preservation (avoiding moisture damage) and indoor comfort (consistent temperatures with no draughts). By focusing on fabric performance, Passivhaus addresses the fundamentals and ensures that these key benefits are realised.

Passivhaus may have been seen as a niche approach and as an interest for a small minority of enthusiasts, but its influence is growing rapidly. Passivhaus simply applies the laws of physics to building for low-energy use and human comfort. These are the principles we all need to understand if we are to construct buildings intelligently. Passivhaus is 'low-energy design' and therefore its principles should be an integral part of any sustainability assessment system, even if the targets lie a little outside the Passivhaus-certified standards, as they do in the CSH. The new Fabric Energy Efficiency Standards (FEES) being considered for inclusion in the 2013 Building Regulations, and already adopted into the CSH, bear the mark of Passivhaus influence. The application of Passivhaus principles to less demanding energy targets, and the relationship to zero carbon, are discussed in Chapter 5.

Architecture has tended to be driven by aesthetics rather than by energy performance. The fact that it has not been uncommon to believe in a future of 'free energy' has meant that our buildings have been relatively 'energy-unconscious'. Once your world view turns towards considering an energy-poorer future, then Passivhaus becomes an essential contribution to addressing this problem, and it challenges us to engage with the building physics and to understand how it all works. Those with an innate fear of science need not despair – once we are prepared to have a closer look, we can grasp the principles, and we should be inspired to build our houses with greater quality and attention to detail; something we have failed to do in the past. We have tended to be rather addicted to immediate short-term solutions; Passivhaus, on the other hand, plans for the longer term – an investment now returns benefits over many decades. Quality ensures longevity.

Passivhaus is not a particular style of architecture – as can be seen from the variety of images in this book. Of course low-energy principles will inform the design, as will many other constraints, such as site location and budget! This is generally the architect's domain, and there is no reason why aesthetic solutions should not arise as readily as in any other circumstances. Certainly design is a skill; truly beautiful and functional buildings are a lifetime challenge. Poor design is a function of our lack of imagination and not a fault arising from restrictions such as low-energy requirements. Restrictions are often what lead us towards elegant solutions.

We hope that self-builders, planners and building control officers, as well as those responsible for formulating and implementing housing policy, will all find this book useful. If low-energy building is to become the norm, the route to achieving this needs more than new legislation. There are important cultural and process issues to be considered, as well as a desperate need for good training and support. In the UK, the government is extremely reluctant to intervene in new markets. But without some supporting and enabling mechanisms, the necessary transition to a low-carbon economy will most likely be slower, and opportunities – both economic and environmental – will be lost.

New Passivhaus houses are being built at competitive rates, and these costs will further improve as the market for low-energy products grows. Our future housing requirements need to be met with economically viable solutions – and Passivhaus, with its focus on fabric efficiency, offers just that.

How to use this book

Chapter 1 is an introduction to Passivhaus construction and an overview of the rest of the book. Chapters 2 to 6 and the final two chapters are fairly broad-ranging, discursive pieces, including some sections that express more personal views, while Chapters 7 to 12 (Part Two) are more practical and technical.

Of necessity, the book does contain some particularly technical content (terms, units of measure and diagrams), which we have tried to make as accessible as possible. Chapters in Part Two also contain explanations of key concepts that need to be grasped by building professionals in order to achieve ultra-low-energy buildings. We hope that readers who are less comfortable with the technical content will nevertheless absorb enough of these chapters to gain an understanding of the principles involved. If you are planning to commission your own Passivhaus, this understanding will help you to select the building team.

We have tried to ensure that each chapter reads as a piece, which means there is, at times, some repetition across chapters. However, it is unlikely that you will want to read the whole book in a single sitting, and there may be particular chapters that interest you and that you may wish to turn to first.

On a very practical level, after reading this book you should be in a much stronger position to brief an architect and/or direct a genuine low-energy project, avoiding many common pitfalls and so reducing risks of time and cost overruns. This book does not offer off-the-shelf 'copy-and-paste' solutions; instead we are aiming to equip you with the tools to create the right solution for your budget and requirements. Where an issue cannot be explained fully, we provide the information for you to recognise when additional support or input is needed and how to approach a building project intelligently to meet its low-energy aims.

We have aimed to avoid listing supplier or manufacturer names, as this information is likely to date quickly due to the rapid developments in both resources and products for Passivhaus, but relevant details are included in the Resources section at the back of the book.

A glossary of terminology, including jargon and specialist terms, is also included. A separate glossary is given of all units. Summaries of the glossary definitions of specialist terms and of units also appear in 'terms explained' boxes in places within the book, wherever such explanations are considered to be helpful. Terms that appear in the glossary, and units, are shown **tinted** the first time they appear, whether in the main text, a table or a 'terms explained' box.

PART ONE

The how and why of Passivhaus

CHAPTER ONE
What is a Passivhaus? 16

CHAPTER TWO
The economics of a Passivhaus 32

CHAPTER THREE
Passivhaus Certification 40

CHAPTER FOUR
Challenges of meeting the Passivhaus standard 46

CHAPTER FIVE
Natural materials, zero carbon and resilience 60

CHAPTER SIX
Setting up a Passivhaus project 74

CHAPTER ONE
What is a Passivhaus?

The Passivhaus energy standard, the Passivhaus concept, common misconceptions, retrofitting and the EnerPHit energy standard

A Passivhaus building is designed to be very comfortable and healthy, and to use vastly less energy than conventional buildings, irrespective of the climate. This is achieved by careful design informed by building physics and, crucially, by thoughtful and careful construction by a properly skilled and motivated team.

Passivhaus originates in Germany; the German word *Passivhaus* literally translates as 'passive house or building', since *Haus* refers to a building as well as a house. The word is being incorporated into English, although the concept is still often referred to as 'passive house', particularly in the United States. When people hear the term 'passive', they sometimes assume that this means no heating system or that the design relies on 'passive solar design', i.e. utilising heat from the sun. There is some truth in these assumptions. A Passivhaus does require a trickle of heat to maintain 20°C, although not nearly enough to justify a central heating system. It relies on high levels of uninterrupted, all-round insulation, airtight design and heat gained from the winter sun through the windows (**solar gain**); however, solar gain is not in itself sufficient to heat a Passivhaus. A Passivhaus is also more comfortable and healthier than a standard build, as there are no draughts, no condensation or mould in cold spots, and the air is fresher. We will see later in the book how and why this is the case.

The thermographic images below, in which surface temperatures are represented with colours, show how high levels of insulation transform a building's energy performance. A well-insulated building in winter has cold external surface temperatures (shown in blue). Thermographic imaging also highlights any '**thermal bridges**' – gaps in insulation that allow heat to bypass it.

Heating and cooling

The Passivhaus concept applies in hot climates as well as in cold: in hot climates the focus is on minimising the energy used to keep a building cool (see box on page 26). Since, at present, Passivhaus is used mainly in cold climates, we refer generally in this book to 'heat used' and 'heat loss'. In a hot climate, this translates to 'energy used' and 'energy loss'.

Thermographic images of the Tevesstrasse project, Germany: before (left) and after (right) its Passivhaus retrofit. *Images: Passivhaus Institut*

Table 1.1 **Passivhaus mandatory technical requirements**

Requirement	Criterion to meet Passivhaus standard	Current (2010) UK Building Regulations* criterion	Average achieved in existing UK housing stock
Airtightness	Below 0.6 air changes per hour (ach)	10m³/hr/m², equivalent to approximately 10ach for a typical home.	More than 10ach
Annual specific space heat (or cooling) demand†	15 kWh/m².a needed to provide space heating to 20°C, or cooling mostly below 25°C in hot climates.	No energy standard is defined in current Building Regulations.‡	Around 200kWh/m².a for heating in the UK[1]
Specific heat load†	10W/m² – the peak power needed to maintain 20°C Internally when it is -10°C outside.	No energy standard has been set.	No data available
Annual specific primary energy demand	120kWh/m².a of primary energy, i.e. the energy consumed at source (e.g. at a natural gas well), needed to meet *all* energy demands.	No energy standard has been set.	Over 400kWh/m².a [2]

* In England and Wales.
† Either one or the other of these two requirements must be met.
‡ But the proposed zero carbon standard planned for 2016 may include a target of 39-46kWh/m².a (equivalent to over 50kWh/m².a in Passivhaus terms because of the stricter definition of usable floor area applied to Passivhaus calculations than that used in current Building Regulations).

Terms explained

air changes per hour (ach) – the measure used by Passivhaus to determine airtightness (the degree of leakage of air from a building): ach is the flow rate of air entering and exiting the building (in m³/hr – metres cubed per hour) divided by the ventilation volume (the total internal air volume, in m³), at 50Pa (pascals) pressure above and below ambient atmospheric pressure.

kWh/m².a – kilowatt hours (kWh – a unit of energy) per square metre [of usable internal floor area, termed 'treated floor area' (TFA)] per annum.

m³/hr/m² – metres cubed [of air entering/exiting] per hour per square metre [of thermal envelope area – the area of floors, walls, windows and roof or ceiling that contains the building's internal warm/heated volume]. m³/hr/m² is the unit of air permeability, the most commonly used measure of airtightness in the UK.

W/m² – watts per square metre [of TFA].

The Passivhaus energy standard

A building is a Passivhaus if it meets a voluntary technical standard that, being international, has to be met regardless of the local climate. It was developed by the **Passivhaus Institut (PHI)**, an independent research institute founded in Germany in 1996. The Passivhaus standard is defined by the core technical requirements listed in Table 1.1 opposite. However, understanding Passivhaus is about a lot more than these numbers. It is a process informed by some key principles, which we will briefly explore next (and in more detail in the rest of the book).

Any building that meets the standards in the second column of Table 1.1 is a Passivhaus. If you don't understand the entire table now, you should become familiar with the ideas as you read the rest of this book. Each requirement is also discussed in the sections that follow. Definitions of terms and units are given in the box below the table, and are also explained in the glossary.

To be sure that a building designed as a Passivhaus genuinely meets the Passivhaus standard and to have independent verification that the design will work as intended, it helps to be certified. Chapter 3 explains what Passivhaus Certification involves and what benefits it brings.

Passivhaus and other building standards

Different standards are not, for the most part, quantitatively comparable, either because they do not measure the same things or, often, because the conventions and assumptions on which they are based are different, making direct comparison impossible. Some are country-specific and stem from government initiatives. Others, such as Passivhaus, originate from non-governmental organisations and are therefore voluntary. Some standards have a narrow focus, while others, such as the UK's **Code for Sustainable Homes (CSH)**, attempt to cover a broad range of sustainability goals. The CSH was originally conceived as an assessment system. By contrast, Passivhaus is not only an energy performance standard but, critically, is also intended to be a design process – which, if applied intelligently, will get you to your low-energy goal.

To some extent, it is possible to mix and match standards. For example, there is no reason why water usage targets from one code cannot be mixed with energy usage standards from another. Some projects are required to meet statutory assessment-based targets, but you can still choose to work to a tougher, voluntary standard. Similarly, there is no reason why other areas of sustainable building not covered by standards or codes cannot be addressed in the design. For example, Passivhaus does not address the properties of building materials – such as **embodied energy** (the energy used in the sourcing, manufacture and transport of a material), breathability and recycle-ability – but clearly there is nothing to stop you choosing to address these independently of any standard, while still achieving the Passivhaus energy-in-use standard. These issues are discussed further in Chapter 5.

The Passivhaus concept

Underlying the Passivhaus concept are several key ideas, discussed on the following pages. While some are 'common sense', others are less obvious and more technical in nature.

Getting the fabric right

The Passivhaus approach concentrates above all on 'getting the fabric right'; in other words, on designing, specifying and constructing the foundation/floor, walls, roof and windows correctly to achieve the Passivhaus standard. Money spent on the building fabric should be seen as an investment for the life of a building (normally a minimum of 60 years), as production of the building materials can consume considerable energy. Designing the fabric to last means that the invested money, energy and carbon provide enduring benefit. In contrast, money spent on 'bolt-on' technologies, such as those needed to provide hot water or space heating (e.g. a boiler or a hot water tank), is a shorter-term investment – 20 or perhaps 30 years at the very most. Those systems may well be replaced several times during a building's lifespan. The smaller the energy burden needed to heat (or, in hot climates, to cool) the building, the less hard the bolt-on technologies have to work, and the smaller and potentially simpler to manufacture, maintain and use they can be.

Optimising the design from day one

The architect's initial designs are modelled for energy performance using the **Passivhaus Planning Package (PHPP)**, specialist software developed by the PHI (see Chapter 7). The PHPP makes it possible to test whether the design will achieve the Passivhaus energy and comfort standard. Once the design has been entered into the PHPP, it is possible to vary elements of it to measure the effect on energy performance. This is an iterative process where client preference, aesthetics, planning considerations, costs and any other practical constraints can be balanced against the design's energy performance. Such an approach brings multiple benefits. Rather than over-engineering the design to make sure it reaches the Passivhaus standard, the design can be optimised. It also means that the client, architect and builder can make informed decisions about the design, that money is not wasted, and that everyone understands the significance of the design and why certain choices were made. Depending on the project, this preliminary detailed work helps to reduce the risk of unexpected cost overruns later in the project.

There is no doubt that this approach is good for the project and makes it easier to reach the Passivhaus standard. However, it is difficult for many to accept the risk of paying for more detailed design work than is typical in a standard build before planning permission has been given. The planning system is discussed in Chapters 4 and 14.

Solar gain and shading

Wintertime solar gain (heat gain from the sun through glazing) is an important part of how a Passivhaus stays warm. However, when a building is designed to minimise heat loss in winter, it is especially important to ensure that summertime solar gain is avoided as far as practicable. This can be achieved by the use of shading devices, which are discussed in Chapter 11.

Insulation and avoidance of thermal bridges

A Passivhaus needs more insulation than other buildings, and that insulation must wrap continuously around the building so that thermal bridges are eliminated (or, in a retrofit, minimised). A thermal bridge, commonly known as a cold bridge, is a gap in insulation that allows heat to 'short-circuit' it. This happens when a material with relatively high conductivity interrupts the insulation layer.

Form factor

The term '**form factor**' essentially refers to the shape of the building: it is the ratio of the external surface area to the internal usable floor area, known as the treated floor area (TFA), and is a measure of how compact or spread-out the design is. A more compact design makes it easier and cheaper to achieve the Passivhaus standard because the walls, roof and floor can be a little thinner (have a higher **U-value**) than would be needed in a more spread-out design. Table 1.2 below demonstrates this.

Thermal comfort

Most discussion of Passivhaus tends to focus on saving energy, but just as central to **Passivhaus methodology** is the concept of **thermal comfort**, defined as the "condition of mind which expresses satisfaction with the thermal environment". More simply, this means not feeling too hot or too cold. In designing a building, the effort to improve thermal comfort needs to be directed at reducing all cold surfaces and draughts. In Passivhaus, there are comfort design criteria (see Table 1.3 overleaf) intended to eliminate all draughts and cold surfaces, and to provide sufficient fresh air. Improving thermal comfort has an added benefit. In a conventional building, we often turn up the heating in cold weather to compensate for draughts and the chilly feeling (known as 'cold radiant') we experience when near cold surfaces. The occupants' need to overcome the effects of draughts and cold radiant often adds to a conventional build's real-world energy consumption. In a Passivhaus, this does not happen, allowing thermostats to be set lower for the same level of thermal comfort.

Table 1.2 **Impact of form factor on U-values needed for a Passivhaus**

Form factor	Typical type of dwelling this might represent	Wall/roof/floor – approximate range of U-values needed to reach below 15kWh/m².a in the UK
<2	Apartment block or terrace (row houses)	0.15W/m²K
2-3	Semi-detached or compact detached two- or three-storey property	0.10-0.15W/m²K
3-4	Less compact detached house or bungalow	0.10W/m²K
>4	Very spread-out bungalow	0.05-0.10W/m²K

Terms explained

U-value – measures the ease with which a material or **building assembly** (a structural part of a building, i.e. walls, floor or roof) allows heat to pass through it; in other words, how good an insulator it is. The lower the U-value, the better the insulator. U-values are measured in W/m²K.

W/m²K – watts per square metre [of the material/assembly in question] per degree **kelvin** [temperature difference between inside and outside the thermal envelope]. (One degree kelvin = one degree Celsius.)

Table 1.3 **Passivhaus comfort design criteria**

Function	Criterion
Ventilation	30m³ per person per hour (needed to provide fresh, healthy indoor air) and maximum air speed 0.1m/s (from supply and extract vents) to avoid draughts and minimise the energy needed to move air around
Minimum internal surface temperature	Greater than 17°C when it is -10°C outside (lower surface temperatures cause convection-driven draughts; this target avoids this, allowing all parts of the building, even those adjacent to windows, to be used)
Overheating limit	Ambient temperatures of above 25°C for fewer than 10% of days annually

Airtightness and indoor air quality (IAQ)

Making a building less 'leaky' is important because air escaping in an uncontrolled way through the building fabric wastes energy, risks reducing the building's lifespan (as a result of air carrying moisture into the fabric) and also makes it feel less comfortable during cold or windy weather. As statutory energy standards improve, there is a trend towards greater airtightness. The problem is that as buildings are made more airtight, **indoor air quality (IAQ)** almost always deteriorates. In some countries, the practice of opening windows daily to ventilate can help a little, but some research (see Chapter 12, page 192) has shown that opening windows provides only a brief improvement of IAQ. In the UK, where buildings were traditionally very draughty, there is no similar custom of regularly purging stale air. New windows often have trickle vents in an attempt to address this problem, but these are arguably ineffective because people are quite often unaware of their existence or significance and they simply remain closed. Even if they are used correctly, the rate at which air is changed is dependent on how windy the conditions are.

A Passivhaus is many times more airtight than typical new builds, so a reliable and consistent method must be used to keep the indoor air fresh and healthy when the windows are closed. Passivhaus uses a carefully designed, highly efficient and quiet heat recovery ventilation system, known as **mechanical ventilation with heat recovery (MVHR)**. In an MVHR system, no air is recirculated, only the heat is recovered and recirculated; and, if correctly designed and installed, there should be no noise or perceptible draught from the movement of air. The result is that IAQ in winter is maintained at a very good level. In the summertime, ventilation can be achieved by leaving windows tilted open and/or using the ventilation system. Chapter 12 explores in more detail how heat recovery ventilation works in a Passivhaus.

Designing correctly for airtightness, or minimal air leakage, is part of the Passivhaus architect's job (see Chapter 9). However, the Passivhaus airtightness standard is a particular challenge for contractors because once the airtightness layer has been created, it is very easy to damage it accidentally later in the build. To avoid this, everyone in the construction team, including subcontractors, needs to understand exactly what the airtightness layer is, why it is important, and what changes in working practice are needed to reliably create and protect it. Chapter 9 explores the challenges contractors face in delivering the airtightness standard on-site.

Annual space heat demand / heat load

These terms refer to the energy and the maximum power needed to heat your home. Achieving either the 15kWh/m².a annual [specific] space heat (or cooling) demand – which represents around 90 per cent less space heating energy than in a typical UK building – or the 10W/m² [specific] heat load is one of the key requirements for a Passivhaus (see Table 1.1, page 18). (Note that, unlike the heat load, the '**cooling load**' requirement for a Passivhaus in a hot climate is not yet fully specified as part of the Passivhaus standard.) Some people ask why these requirements aren't made even lower. Why not completely design out the need for any space heating or cooling? While this is technically possible, it significantly increases the build costs for relatively little additional energy saving. The choice of 15kWh/m².a represents an optimum point where the need for conventional central heating (or, in hot climates, air conditioning) is eliminated. Space heating options in a Passivhaus are discussed overleaf.

Annual primary energy demand

The 120kWh/m².a annual [specific] primary energy demand requirement (see Table 1.1) is designed to ensure that consumption of energy for hot water, cooking, appliances, lighting and all other uses is efficient. It effectively makes it impossible to use **internal heat gains** from inefficient appliances or poorly designed hot-water pipework as a tactic to get around the 15kWh/m².a space heating requirement. It also discourages the use of electricity for direct water heating (for instance, an immersion heater) and encourages the use of solar hot water where practicable. Eliminating excess heat gain from inefficient domestic hot water systems and appliances also has the benefit of reducing the risk of your house overheating in summer.

Working as a team – building trust between architect, builder and client

Working as a team may well sound like a clichéd aspiration, but it is an ideal we fall short of in many real-world builds. Being able to achieve an effective, trusting and cooperative working relationship between architect, builder and client is critical to the success of a Passivhaus build. In the same way that a Passivhaus build demands much attention to technical details, this 'human' detail is one that must also be addressed.

Typically, architects work on their designs with minimal input from those who will be tasked with building them. While this approach does not preclude a close, trusting relationship between the key parties, it does make it harder to achieve in practice. If the builder is involved in the initial stages of the project and has a meaningful input into the construction details, he or she should be able to help reduce the complexity of the build without affecting the aesthetic or the function. More importantly, early involvement of the builder will help to build trust and mutual respect with the architect. The builder is more likely to accept the design and to have a deeper understanding of its rationale.

Common misconceptions about Passivhaus buildings

While it is fairly easy to grasp what a Passivhaus is about, it remains an abstract matter for those of us who have never experienced one, and so most people find it hard to predict how they would feel about living in such a place. We have long experience of buildings that are draughty and hard to heat, or stuffy and tricky or expensive to keep cool. Many of the misconceptions

Heating options in a Passivhaus

One of the aims of Passivhaus is to simplify the technology needed to provide a comfortable indoor environment. And one of the challenges of Passivhaus is to keep this in mind while sourcing heating from what is currently available on the market. Most heating options are designed for less efficient buildings and are therefore often oversized (they deliver too much heat) and sometimes overly complex.

Although very little space heating is needed, without any heating most occupants would find the internal temperature too cool. For an average-sized UK home built to the Passivhaus standard, the 10W/m² specific heat load translates into a total heat load of less than 1kW. A similar-sized standard-build home would typically be fitted with a 24kW natural gas boiler (admittedly, most heating engineers tend to oversize the boilers they install). This means that an average-sized Passivhaus could be heated with a single radiator in the main living space and heated towel rails in the bathroom(s).

The Passivhaus standard does not explicitly specify what form of heating should be used. As we saw on the previous page, the annual primary energy demand limit discourages the direct use of electricity for water heating. This limit similarly discourages the direct use of electricity for space heating. The specific heat load limit of 10W/m² is low enough to allow heat to be provided via the ventilation system. A small **supply duct radiator** adds heat (via hot water or an electric element) into the supplied air (see Chapter 12, page 202). Allied to this, in average and smaller-sized dwellings, the PHI encourages the use of '**compact units**', which combine the ventilation function with the provision of hot water and the input of the small amounts of heat, distributed via the ventilation system, needed for space heating.

Where a Passivhaus apartment block or a group of detached buildings is being planned, it is possible to use district heating, whereby heat is produced centrally and distributed via insulated pipes. Each dwelling can still control the amount of heat used and individual usage can be measured using heat meters. This approach has two advantages: first, rather than having multiple boilers to maintain, there are only one or two; second, a broad range of larger units is already available on the mainstream market.

It is worth remembering too that in a Passivhaus the heating season is shorter than in standard builds, because during the 'shoulder' months (October, and February to April) there is enough solar gain to maintain a comfortable temperature without heating. This means that, in addition to lower energy use, the lifespan of the heating technology should be extended because it is used less.

Some of the main advantages and disadvantages of the different heating and hot water options in a Passivhaus (aside from via the MVHR) are set out in Appendix A. Unfortunately, there are no options without some drawbacks. Whichever solution you settle on, it does need designing by someone with adequate knowledge of the system or product. Ultimately the choice of heating system will depend on many variables, not least budget! Those concerned with sustainability issues, particularly our climate predicament, may be attracted to wood as a fuel source. This is discussed in Chapter 5, and mentioned briefly on pages 26-7.

about Passivhaus reflect this. Here are a few of the most common.

"You are not allowed to open the windows / you are locked away from nature"

In summer, a Passivhaus relies on windows that can be opened. In warm weather, the most effective way of purging any heat built up during the day is to open the windows at night-time. During the winter heating season, there is no reason why a window can't be opened, but most people will feel less need to do so than in a standard build because the MVHR system ensures good air quality. A few people who previously always slept with an open window in summer and winter continue to do so in their Passivhaus home. This reduces the temperature (in winter) slightly and adds a little to the building's energy use, but the Passivhaus will still function as intended.

"The air is too dry in a Passivhaus"

This can be an issue in Passivhaus designs, but it is also true of conventional centrally heated buildings. However, the problem can be minimised, provided that the MVHR system is specified, installed and commissioned correctly. Dry indoor air is caused in the wintertime because cold winter air holds a lot less water vapour (moisture) than air at typical room temperature. Even if the winter air is completely saturated with water vapour (i.e. it has a 100-per-cent **relative humidity (RH)**), as happens on a cold, rainy winter day, it still holds a lot less moisture than indoor air at 20°C, even if that indoor air is only at 50 per cent RH. This means that dry winter air entering the building, whether controlled via an MVHR system or uncontrolled through leakage gaps in the building's structure, will feel very dry (i.e. have very low RH) as the air reaches room temperature.

Dry wintertime indoor air is more of a problem in the colder and drier winter climate of central Europe. In warmer, damper winter climates, such as those of the UK and Ireland, the moisture content of dry indoor air can be increased by, for example, keeping indoor plants and drying clothes on a washing stand – as well as by normal cooking, washing and bathing activities, and by the occupants' breathing. Use of particularly **hygroscopic** materials (those that act as a water vapour buffer by absorbing and releasing a lot of atmospheric moisture), such as clay or hemp, can also help to moderate fluctuations in internal RH. There are MVHR units that recover humidity as well as heat ('ERV' units, which are used in the USA), but this tends to be at the expense of efficiency.

"A Passivhaus overheats in summer"

With appropriate design strategies, this should not be a problem. As mentioned, it is essential to be able to open windows at night during warm spells. In the daytime, windows exposed to summer sunlight should be adequately shaded externally (see Chapter 11). The Passivhaus standard requires that the internal temperature does not exceed 25°C for more than 10 per cent of days annually, although we suggest aiming for a more stringent target of 3 to 5 per cent – this should not require much additional shading and will make a big difference to comfort levels in summer. In hotter, tropical and subtropical climates (see box overleaf), the effort to keep within the 15kWh/m².a requirement focuses on the annual *cooling* rather than *heating* demand.

"There is no heat source such as a cosy fire"

In a Passivhaus there is no physical need to light a fire in order to feel cosy, for the same reason that we don't feel the need to light a fire in a

This Passivhaus in Louisiana illustrates design strategies for a hot climate. *Image: Catherine Guidry*

Passivhaus in hot climates

In hot climates, the design solutions to meet the cooling demand (i.e. energy used for cooling) will in many respects be different from those adopted in colder climates, although the principles of airtightness, thermal-bridge-free construction and high levels of insulation still apply, to protect the building from overheating. Effective external shading is essential to avoid solar gain. The ventilation system needs to include energy recovery to both pre-cool and, in humid climates, dehumidify the supply air. (Humidity and indoor air quality are discussed in Chapter 10.)

The design features used in the Passivhaus pictured here, in southern USA, include:

- large external shading devices to counteract heat gains all year round
- rain screen as a shading device for the walls, with a radiant barrier behind to reflect heat
- reflective, light-coloured metal roofing
- low **g-value** windows (see Chapter 11).

conventional house during temperate summertime weather (where external temperatures are similar to room temperatures). It is possible to install a very small wood burner in a Passivhaus, but it needs to have a very low output for space heating – typically less than 2kW. Using it on all but the coldest winter days could overheat the house, unless you were very careful about how much wood was added (although this could easily be remedied by opening windows). In a Passivhaus, a wood burner needs its own combustion air supply from outside and exhaust to outside. Also, the air supply and exhaust must be detailed correctly where they enter and exit the building, to avoid thermal bridging (see Chapter 8) and optimise airtightness (see Chapter 9).

Suitable wood burners that deliver very little space heating are coming on to the market, so this could be an option if having a log fire is a priority (and if there are no constraints on wood supplies). Otherwise, some log burners have a marble or granite surround, which acts as a **thermal mass** – absorbing and storing heat, and so limiting the rate at which the fire's heat is released. Alternatively, there is no reason why a home cannot be designed with some areas that are outside the Passivhaus thermal envelope. A standard log burner or even an open fire could be used to heat those spaces.

It is quite possible to use a heat source that burns woodchips or wood pellets. The challenge when designing a single domestic-scale Passivhaus building is that such boilers have tended to have higher heat outputs than are needed. But in multi-home Passivhaus projects a single wood-fuelled boiler could be used to serve a number of homes. Chapter 5 explores the options for making a Passivhaus **zero carbon** (no net greenhouse gas emissions), including using wood as a fuel.

"The ventilation system is noisy, overly complex technology that uses energy to run and transmits noise between rooms"

A Passivhaus-certified MVHR unit, correctly designed and installed in a building designed and constructed to Passivhaus standards, is not noisy and saves much more energy than it uses. An MVHR works entirely differently from air conditioning, with which it is sometimes confused. With heat recovery ventilation, fresh air from outside is slowly introduced to parts of the house (typically the sitting room and bedrooms), each room being supplied separately via its own duct. From these rooms the air moves through common areas such as halls and stairwells, and is extracted from the kitchen and bathrooms. The rate of air movement in an MVHR system is a lot slower than in an air conditioning system, making it quiet and energy-thrifty. Sound attenuators (soft cylindrical inserts) are placed if necessary at specific points in the ducting system to muffle any residual noise from the MVHR. Chapter 12 looks at ventilation in a Passivhaus in detail.

"A Passivhaus can't be built with natural materials or 'low-impact' materials"

There is no reason why a Passivhaus cannot be constructed using low-embodied-energy and low-impact materials (building materials that do not require large energy inputs to make and deliver to site, and which can be reused, recycled or otherwise returned to the environment without significant impact or cost to the environment). Such materials are discussed in Chapter 5. The Passivhaus standard has a narrow focus on energy consumed in the use of the building, and the Passivhaus Institut has deliberately chosen not to stipulate any standards for embodied energy or other environmental impacts of the materials chosen. This is simply because the PHI does not want Passivhaus to be seen as an eco-niche solution. Passivhaus will make an impact on energy use and carbon emissions only if it is put into practice as widely as possible. This approach is showing dividends in Germany and Austria, where public authorities are moving towards making Passivhaus the norm for new building.

"Passivhaus architecture is boxy and unimaginative, not joyous or uplifting"

Passivhaus is not intended to be a single architectural style, although it may sometimes be seen as such because so many early examples were built in a relatively small part of central Europe. It is likely that the designers of those early Passivhaus buildings drew their inspiration from the region's existing architectural legacy and practice.

More recent examples in North America and the UK are beginning to change this assumption, as can be seen from the buildings pictured overleaf. In 2010 the PHI started an awards scheme for

Different styles of Passivhaus buildings. **Above:** SurPlus-Home, USA. Winner of the 2009 Solar Decathlon. *Image: ENERGATE®* **Top left:** The first home in the Channel Islands to be Passivhaus-certified. *Image: David Aslett* **Left:** Family home in Chemnitz, Germany. *Image: Passivhaus Institut* **Bottom:** Centre for Disability Studies, Essex. *Image: Simmonds.Mills*

good Passivhaus design, since it is keen to encourage design diversity: as the PHI sees it, Passivhaus needs to adapt and adopt the design traditions of the societies where the buildings are constructed, not least because those traditions were born of years of practical experience of building in very different climates, with different local materials and skill sets.

Passivhaus design tends to encourage the creation of a fairly compact thermal envelope, as we noted earlier (page 21). This does not prevent you from building a Passivhaus that is very spread out – but doing so makes it more expensive to achieve the Passivhaus standard, as **building elements** (materials or objects that are part of the structure of a building) need to be significantly thicker to achieve the lower U-values needed in a Passivhaus. In practice, economics, local architectural tradition and the constraints of the local planning system are much more significant determiners of building shape.

It could be said that it has always been part of an architect's remit to design buildings that are 'joyous' or 'uplifting'. Architects work with many constraints: climate, project budget, planning laws and **Building Regulations**, space- and site-specific restrictions, the skill set of the builders, and so on. Clearly, it would be untrue to say that Passivhaus uniquely inhibits architects from designing joyous buildings. The requirements of Passivhaus do influence appearance, but this can be handled sensitively or clumsily, depending on the skill and imagination applied. New aesthetics often take time to filter into public appreciation (we tend to warm to familiar features), so this is not a definitive test for aesthetic value.

We have lived through a period when energy was cheap and its use (apparently) consequence-free; when it was feasible to build very energy-hungry structures. The world is changing and energy consumption is becoming a real constraint for growing numbers of people. The accepted aesthetic will therefore be challenged, but alternative aesthetic solutions do evolve.

What is the significance of refurbishment or retrofit?

Most of the housing stock that will exist in 2050 (the year widely referred to as the target for 80-per-cent reductions in carbon emissions) already exists today. In the UK, new build replaces only a tiny proportion (around 1 per cent) of the housing stock annually. This replacement rate is constrained by economics and by land ownership patterns, planning law and culture. Because of this, most energy / carbon reduction in the housing sector will come from retrofitting existing building stock.

No agreed standard definition of refurbishment or retrofit exists, but in the UK it generally refers to fairly superficial changes, such as new kitchen, bathroom or decorative changes, rather than addressing any backlog in maintenance of the existing fabric or services. However, a refurbishment / retrofit nearly always provides an opportunity to improve the building's energy performance at much lower additional (marginal) cost. It pays to plan strategically, so that no work completed in earlier phases has to be undone later. That way, it is possible to make substantial and cost-effective improvements over time, in stages, as resources allow. The terms 'refurbishment' and 'retrofit' are often used interchangeably, but since 'refurbish' implies superficial changes as opposed to changes to a building's fabric, in this book we will use the term 'retrofit' to describe the adaptation of existing buildings to **ultra-low-energy** standards.

Passivhaus and retrofitting

A minority of Certified Passivhaus projects have been retrofits of existing buildings. However, in practice, reaching the full Passivhaus standard – in particular the 15kWh/m².a space heat demand and 0.6ach airtightness limits – is not often feasible. This issue is explored in Chapter 6.

During 2010 and early 2011, the PHI piloted a new energy performance standard for residential retrofits, known as **EnerPHit**.[3] This was launched at the Passivhaus Conference in May 2011. EnerPHit allows a maximum annual heat demand of 25kWh/m².a and an upper airtightness limit of 1.0ach, if the 0.6ach target can be shown to be impracticable. EnerPHit also sets requirements for individual elements of a retrofit, should the 25kWh/m².a requirement not be met. The EnerPHit standard is therefore more complex than the Passivhaus standard. As of November 2011, only retrofits in certain climates, including central Europe and the UK, can be certified to the EnerPHit standard. The key points of the EnerPHit standard are:

- annual specific space heat demand – below 25kWh/m².a
- airtightness – normally 0.6ach; if this is demonstrated to be unachievable, then up to 1.0ach with additional evidence provided
- evidence and calculations to demonstrate that moisture management issues have been adequately addressed
- only existing buildings that are unable to meet the full Passivhaus standard are eligible for EnerPHit certification.

If the 25kWh/m².a limit is exceeded, EnerPHit certification is also possible by meeting criteria for individual components and building elements:

- **walls** – externally insulated wall (>75% of total wall area), U-values below 0.15W/m²K; internally insulated wall (<25% of total wall area), U-values below 0.35W/m²K
- **roofs / top-floor ceilings** – U-values at or below 0.12W/m²K
- **floors** – U-values at or below 0.15W/m²K
- **windows** – installed whole window, U-values at or below 0.85W/m²K
- **external doors** – installed whole door, U-values at or below 0.80W/m²K
- **linear thermal bridges** – at or below +0.01W/mK; point thermal bridges +0.04W/K
- **ventilation** – heat recovery efficiency at or above 75%; electrical efficiency at or below 0.45Wh/m³.

While retrofit is more challenging than new build, the EnerPHit standard provides an achievable

Terms explained

thermal bridge – a gap in insulation that allows heat to 'short-circuit' or bypass it. This occurs when a material with relatively high conductivity interrupts or penetrates the insulation layer. Thermal bridges occur in 'point' or 'linear' form (see Chapter 8).

U-value – measures the ease with which a material or building assembly allows heat to pass through it, i.e. how good an insulator it is. The lower the U-value, the better the insulator. U-values are measured in W/m²K.

Wh/m³ – watt hours per cubic metre. Used by the Passivhaus Institut to measure the electrical efficiency of MVHR units.

W/K – watts per degree kelvin [temperature difference between inside and outside the thermal envelope].

W/mK – watts per metre [length of the linear thermal bridge] per degree kelvin [temperature difference between inside and outside the thermal envelope].

W/m²K – watts per square metre [of the material/assembly] per degree kelvin [temperature difference between inside and outside the thermal envelope].

benchmark that encourages the same high standards, rigorous methodology and attention to detail that Passivhaus demands for new builds. It is worth remembering that, at 25kWh/m².a, the EnerPHit energy performance standard is still about twice as demanding as that of the **Fabric Energy Efficiency Standard (FEES)** defined in the UK 2010 *Code for Sustainable Homes: Technical Guide*.[4] Similarly, the EnerPHit minimum airtightness standard is about three times as demanding as the zero carbon (CSH Level 5/Level 6) standard. As such, achieving the EnerPHit standard remains a big challenge for architects and builders. However, it is easier to achieve in the mild climate of, say, south-west England than in central Europe, where it was piloted.

RECAP

Passivhaus is all about carefully designing and constructing buildings to use vastly less energy (the reductions are by an order of magnitude) than conventional buildings.

Passivhaus originated in Germany in the early 1990s, but has now spread around the world: it is truly an international standard. It has a deliberately narrow focus on getting the building fabric right in order to achieve low energy in use. While Passivhaus addresses only this aspect of sustainable design, there is nothing to stop house builders who want to build ecologically or sustainably from using a combination of the Passivhaus standard and other standards for, say, water use, or indeed other non-codified environmental objectives, in their specification.

Achieving Passivhaus is not about lots of 'advanced' technology; rather, it is about changing the way we build. This means integrating design for low energy into the plans from day one, designing with an awareness of the impact of form factor, eliminating thermal bridges and radically reducing air leakage compared with standard builds, incorporating additional insulation and making use of solar gain. Once the project goes on-site, the build team needs to work in a more tight-knit, cooperative and mutually trusting way than is currently common in building projects, in order to avoid abortive work and unnecessary costs.

There are many misconceptions about Passivhaus – for example, that one is not allowed to open any windows. The discourse about Passivhaus sometimes results in the tendency to create false dichotomies, in particular between Passivhaus requirements and the use of natural or 'low-impact' building materials. In time, as more and more Passivhaus buildings are completed, it will become easier to overcome such concerns.

Passivhaus can also be applied to retrofitting of the existing housing stock. A newer, slightly less demanding standard specifically for retrofits – EnerPHit – has been devised by the Passivhaus Institut.

CHAPTER TWO

The economics of a Passivhaus

Economic culture, energy costs and supplies, factors affecting property prices, factors affecting Passivhaus costs, methods of determining cost-effectiveness

Chapter Two • The economics of a Passivhaus

Economics is an uncomfortable topic for many people. Unfortunately, few of us can afford to ignore money constraints in a build project – particularly now, in the post-2008 credit-crunch era. As the financial system retrenches in an attempt to recover stability, access to capital has become much more restricted and expensive. The brief period of spending significant additional sums purely to 'save the planet' seems to have long gone. However, counterbalancing this is the growing impact of persistently high, and increasing, energy prices. A few years ago, the energy costs of a home probably didn't figure on most people's agendas, but for growing numbers of buyers this is now becoming a significant factor.

Culturally, we in the UK seem to be somewhat averse to capital spending (one-off spending on durable items or infrastructure) – even if not making the investment means much higher ongoing costs and poorer performance. This was probably true even during the years of relatively strong economic growth up until 2008, when borrowing was easy and cheap. We also appear to have lost the ability to make strategic, long-term decisions to commit to capital-intensive infrastructure projects that would bring medium- and long-term benefits. In the construction sector, where there is a choice between spending the minimum capital outlay needed to meet regulatory and legal obligations – irrespective of the impact this has on ongoing running costs – and spending a few per cent more to achieve hugely lower running costs and other quality benefits, the choice made is usually to spend the minimum. At the same time, many willingly increase their build costs to fund a luxury, high-status or high-profile item, such as a luxury kitchen, without any attempt to assess its cost-effectiveness or payback. Clearly, as house builders and purchasers, we do not act as purely rational economic players.

Energy costs and supplies

In a world of apparently never-ending, cheap energy delivered without environmental impacts, the economic arguments for Passivhaus would be weaker. Most of us in the rich world have got used to living with easy, affordable energy and find it hard to conceive of a different reality. However, we are at a turning point in our history. After 200 years of low-cost energy that could be produced year-on-year at ever-greater flow rates, we are facing a period where energy is much more expensive and where energy supply rates shrink year-on-year, constrained by physical limits to the rate at which energy can be extracted from our environment[1] (this is discussed further in Chapter 5). There is a strong correlation between global economic growth and growth in energy flow rates.[2] Ultimately, it will not be possible for the overall level of economic activity to increase globally if the energy supply is decreasing. This most profound change in our society will change the economics of Passivhaus (and many other things!), making ultra-low-energy buildings extremely attractive. The higher cost of energy, and supply scarcity issues, will change the economics of all building to discourage indiscriminate use of building materials that have a high embodied energy or have travelled long distances.

Macro factors affecting property prices

In Chapters 6 and 14, we touch on the some of the factors that influence the cost of a build – for example, the effects of planning requirements and VAT policy. However, the cost of buying a home or business premises is determined most significantly by the following factors.

- Land prices, particularly in the most densely populated and economically vibrant areas. In

the UK, land prices are comparatively high owing to the concentration of land ownership, the practice of 'land banking' by large building developers (where land on the edge of towns is bought up and held in anticipation of a future change in planning status), green belt legislation and other planning controls on development. This particularly impacts on the cost and availability of building plots for self-builders and small-scale developers.
- Changes in the average number of occupants per household and a growth in UK population. Until the 2008 downturn, the trend has been towards fewer occupants per household. This trend increased demand for homes in a market with relatively slow supply growth, thus increasing house prices.

Overlying these are the large cyclical property price fluctuations experienced by the UK and many other countries (see Figure 2.1). The impact of all these factors dwarfs any price differences between Passivhaus and a standard build.

Factors affecting Passivhaus costs

As we will see later in this book, there are costs and savings involved when building a Passivhaus, compared with the cost of an equivalent standard build that just meets current Building Regulations. The additional cost of building a Passivhaus has been estimated by the Passivhaus Institut[3] at between 3 and 8 per cent, and by the Passive-On Project[4] (having examined costs across a number of European countries) at between 3 and 10 per cent. However, the trend is towards cost parity. This is because:

- building codes are setting tougher energy performance standards

Figure 2.1 Historical UK house prices. Prices are adjusted for inflation, based on 2012 pounds. *Source: Nationwide Building Society, www.nationwide.co.uk/hpi/downloads/UK_house_prices_adjusted_for_inflation.xls*

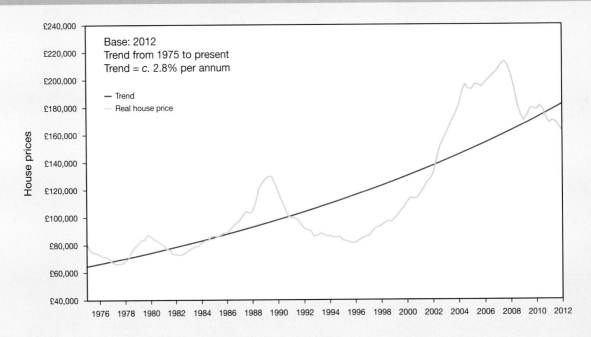

- knowledge and experience of successfully building Passivhaus is expanding
- products associated with Passivhaus are becoming more mainstream.

The following sections consider the effect on costs of different build elements.

Ventilation and heating

In a Passivhaus, where the ventilation system can also be used to supply heat, the cost of a heat recovery ventilation system – the Passivhaus-certified MVHR unit, ducting and other components as well as commissioning costs – can be offset against the money saved by not having to provide a full central heating system and extract vents for the kitchen and bathrooms. Even if the MVHR system is not used to supply heat, whatever heat source is used can be simple and very small-scale – a low-powered wood burner, perhaps. The trade-off is also true in an apartment block, as the economies of scale that apply to conventional communal heating systems also apply to their Passivhaus equivalents.

Insulation

A Passivhaus requires more insulation than a standard build. The Passivhaus Institut has estimated[5] extra insulation costs of between €0.40 (30p) and €1.20 (£1) per 10mm per m² of opaque building element area. As we saw in Chapter 1 (Table 1.2, see page 21), buildings with an efficient form factor require less insulation – the U-values can be higher – in order to reach the Passivhaus standard, so are cheaper to construct.

Where practicable, using insulation that is quick to install on-site or is pre-installed should help keep costs down. Board-based insulation has to be cut very accurately to avoid thermal bridging. This takes more skill and time, adding to costs. **Blown-in** ('pumped in') insulation such as Warmcel® (cellulose) can also be cost-effective, as it requires very little labour to install and there is no waste.

Thermal bridges

As we discuss in Chapter 8, managing thermal bridges at building element junctions is primarily a question of detailed design and careful, accurate execution. An architect or building designer new to Passivhaus will need to spend additional time finding solutions that design out thermal bridging, but this extra cost diminishes significantly as experience is gained in how to do this. Similarly, a builder new to Passivhaus may need to take longer to ensure that unfamiliar details are built correctly. The junctions between the base of the external walls and the floor are particularly challenging, as they have to be constructed to carry much of the weight of the building. This often means using high-cost insulation that has a high compressive strength, such as Foamglas® or similar. In larger buildings such as apartment blocks, completely eliminating the floor–wall thermal bridge may not be as cost-effective as increasing insulation thicknesses – which will be thinner anyway because an apartment block has a very efficient form factor (see page 21).

Airtightness

In a new build, airtightness should add very little to the construction cost of a building, provided that sufficient thought has gone into its design, planning and execution. Again, if the architect or designer and builder are new to Passivhaus, they will need to spend more time to be sure of achieving below 0.6ach (air changes per hour). Costs will start to escalate if mistakes on-site require corrective work. In a well-conceived and well-executed design, the cost of airtightness tape, grommits and membrane will be very low – less than £1,000 in a typical domestic-scale build, unless used excessively (as in the photo below!).

In a retrofit, the challenge of reaching below 0.6ach is much greater, and there will be additional costs associated with it. In the Totnes Passivhaus, which achieved an airtightness of 0.2ach, significant additional work was needed to ensure that there was no gap in the airtightness layer around the existing joist ends (see also Chapter 9, page 139).

Excessive taping!

Making existing joist ends airtight: plastering and applying airtightness tape.

Windows

Passivhaus-suitable and Passivhaus-certified windows are more expensive than standard double-glazed windows, which just meet current (2010) Building Regulations. At present, a window suitable for a Passivhaus is around two or more times the cost of a standard window. At first glance, downgrading the windows may seem like an easy way to cut down the cost of the build: if the U-values of the other cheaper elements of the building are uprated, the windows can be downrated to standard double glazing and the building may still make the 15kWh/m².a space heating demand. However, as we will see in Chapters 7 and 11, in cool-temperate European climates there are good reasons for the use of triple-glazed windows with warm-edge **spacers** (the dividing strip along the edge of a double- or triple-glazed unit that separates each pane) and other specific characteristics. In cold weather, use of standard double-glazed windows would create an internal surface cold enough to reduce thermal comfort because of the cold radiant from the internal window surface. The cold surface would also create convection draughts. Both these phenomena would require a warmer ambient temperature to regain the same level of thermal comfort.

At this point in time, there is no way to avoid paying extra for Passivhaus windows, although the price difference between Passivhaus windows and standard windows will shrink as building energy performance standards are tightened and Passivhaus becomes more mainstream, thereby making windows with an energy performance suitable for Passivhaus much more common.

One way to reduce the cost of windows is to moderate the glazing area, using a mixture of fixed and opening windows (fixed windows are cheaper) and choosing fewer reasonably large (but not too large) windows, rather than more smaller ones. Windows that are part of a manufacturer's standard range, made in standard colours and finishes, will also be cheaper. Although most of us tend to want the flexibility of being able to open any windows, and many may have concerns that with fewer windows there will be insufficient daylighting, costs can be managed by being pragmatic.

The other factor to consider is the windows' longevity. In the UK, most window companies offer products that, while cheaper than Passivhaus-suitable products, are quite often not very durable. Passivhaus windows are generally designed for a much longer service life. When this is taken into account, the annual cost of use will be quite favourable; possibly lower than that of standard windows.

Construction

An architect who also has experience as a builder, or who works with or gets professional input from an experienced builder (particularly one with Passivhaus experience), should be able to devise solutions that are more cost-effective to execute. By optimising the build process it should be possible to claw back enough of the budget to fund most, if not all, of the additional cost of building a Passivhaus. We hope and expect to see examples of this emerging in the near future.

Methods of determining cost-effectiveness

Some assume that Passivhaus is only for the wealthy. In fact the Passivhaus standard was developed with modest housing application in mind, not just for bespoke 'grand designs'. And the whole-lifetime cost of a Passivhaus is considerably lower than that of a standard build,[6] which is why many who adopted the model early on were those with a long-term interest in their property – for example, registered social landlords (RSLs). When you start to plug in probable energy cost increases and energy scarcity factors, the cost benefits of Passivhaus are even greater. But, assuming for now that building a Passivhaus is a few per cent more expensive than a standard build, what is the best way of measuring the cost-effectiveness of the additional investment?

Payback period

'Yes, but what is the payback period?' is the oft-heard refrain of the sceptic. 'Payback period' is regularly referred to in the media, probably because using it confers economic credibility on the speaker (to the lay ear at least) and because 'payback period' is easy to understand: if a build costs £x more and saves £y more per year, simple arithmetic will tell you how long it will take to break even. However, this is not a very useful measure of cost-effectiveness, because the length of a payback period (which is, of course, an estimate) is quite sensitive to small changes to the assumed rate of increase in energy prices and associated costs (such as maintaining the larger-scale heating infrastructure in a standard house compared with its leaner, simpler counterpart in a Passivhaus). Depending on the energy price assumptions used in the calculation, payback period can be used to put a case for or against any given investment.

The fairly conservative example in Table 2.1 overleaf compares a 100m² Passivhaus with a reasonably energy-efficient new 100m² home built by mass-market housing developers. Even with a relatively modest 7 per cent annual real-terms increase in energy prices, the cumulative savings over 25 years – the length of a typical mortgage – on energy costs alone grow dramatically. These figures exclude the costs of

Table 2.1 Cumulative savings on energy costs in a Passivhaus compared with a standard new build over 25 years, at 7% energy price inflation

Year	Energy costs (£/kWh)	Cumulative energy costs		Cumulative savings
		100m² Passivhaus (15kWh/m².a)	100m² standard new build (90kWh/m².a)	
1	£0.05	£75	£450	£375
2	£0.05	£155	£932	£776
3	£0.06	£241	£1,447	£1,206
4	£0.06	£333	£1,998	£1,665
5	£0.07	£431	£2,588	£2,157
6	£0.07	£536	£3,219	£2,682
7	£0.08	£649	£3,894	£3,245
8	£0.08	£769	£4,617	£3,847
9	£0.09	£898	£5,390	£4,492
10	£0.09	£1.036	£6,217	£5,181
15	£0.13	£1,885	£11,308	£9,423
20	£0.18	£3,075	£18,448	£15,373
25	£0.25	£4,744	£28,462	£23,718

maintenance and replacement of heating infrastructure (costs of which are lower in a Passivhaus, simply because less heating infrastructure is needed).

A 100m² home might have a gross area (the area used for costing purposes) of 125m². At £1,200/m², the construction cost would be £150,000. A similar structure built as a Passivhaus, with a 5-per-cent cost uplift, will add £7,500 to the price. Factor in the reduced costs of heating system maintenance and the likely resale advantage, and, clearly, all but a client with the most short-term interest would save money. And, of course, a Passivhaus is built with a much longer design life than 25 years. The savings start to ramp up rapidly in the following decades, even if very conservative assumptions are made about future energy price increases. While we cannot predict future energy prices with absolute certainty, the weight of evidence and an examination of the current trends clearly indicate that we are highly unlikely to return to a world of cheap and easy energy.

Capital costs versus energy costs

Another way to look at the cost-effectiveness of investments is to compare two costs: 1) the extra annual repayment costs on a mortgage that result from an extra capital spend (Figure 2.2 opposite shows how to calculate this); 2) the savings in energy costs and in associated heating system maintenance and replacement costs arising from that extra capital spend. The advantage of this method is that it can be used to inform individuals' decisions about how far and where to direct marginal capital investment to best effect. Once the building design has been modelled in the Passivhaus Planning Package (PHPP), elements can be varied according to how the additional capital investment is spent. This translates into a figure for the number of kWh per year saved for a given option, which can then be converted to a financial value. Maintenance and replacement costs would have to be calculated separately.

While 'payback period' can support a more polemic argument on costs, 'capital costs versus energy costs' is a more useful tool for assessing the cost-effectiveness of a capital investment. This method is used by the Passivhaus Institut to demonstrate the economic argument for Passivhaus: its main benefit being that it is much less sensitive than payback period to assumptions made about energy price inflation. Given that such assumptions are never precise, this sensitivity problem makes 'payback period' a less useful analytical tool. As noted, it is more useful as a tool to support a political view.

$$A = \frac{\text{Extra capital spend}}{\left(\frac{1 - (1+p)^{-n}}{p} \right)}$$

A = Annual repayment costs
n = Number of years of interest payments (length of mortgage or loan)
p = Interest rate (percentages expressed as numbers: e.g. 0.04 for 4%)

Figure 2.2 Calculating the ongoing annual cost of a capital outlay.

RECAP

While the up-front costs of building a Passivhaus are currently around 3 to 10 per cent more than for an equivalent standard building, the trend is towards cost parity, as Passivhaus knowledge and skills become more mainstream and experience is gained by architects and builders. And when the whole-life running costs of the building are included in the costing, a Passivhaus is cheaper than a standard build. Buyers are becoming more aware of energy running costs when choosing a home, so a Passivhaus, with its bullet-proof protection from fuel poverty and its high levels of interior comfort, should represent an opportunity for innovative developers.

In the UK, other factors have a bigger impact on house prices than the extra build costs associated with a Passivhaus. In particular, the concentration of land ownership, 'land banking' and planning constraints have resulted in scarcity of, and very high prices for, building plots, particularly for self-builders and small-scale developers.

Rising energy prices and a growing risk of interruptions in the energy supply will put upward pressure on all building costs (whether Passivhaus or not). The cost-effectiveness of a Passivhaus build can be assessed in different ways, but a method that is less sensitive to assumptions about future energy prices is the most meaningful.

CHAPTER THREE
Passivhaus Certification

Certification of buildings, building components, designers and contractors; the process and information required for building certification

Passivhaus Certification is an independent quality-control process designed to ensure that the Passivhaus standard is reliably achieved. A Passivhaus Certificate can apply to entire buildings, building components (including very low thermal bridge components) and building designers and contractors.

Certification of buildings

In order to gain Passivhaus Certification of a complete building, you need to follow a quality-assurance process applied to its design and construction.

Who certifies a Passivhaus?

Certification of buildings can be carried out only by Passivhaus Certifiers accredited by the Passivhaus Institut in Darmstadt, Germany: there are more than 30 of these worldwide. As of summer 2012, there are three Certifiers in the UK: **BRE**, Cocreate and WARM: Low Energy Building Practice; and one in Ireland: MosArt (see Resources). A Passivhaus project cannot use the same person to act both as Passivhaus Designer/Consultant and Certifier.

Why certify a building?

Passivhaus Certification is all about quality assurance and the avoidance of 'greenwash' – the tendency to make unsubstantiated claims about environmental or energy performance. By choosing to get a building certified, those working on a project have access to expertise and advice that will save money and time, particularly if they have little or no previous experience of trying to meet the Passivhaus standard. During the project, many assumptions have to be made, and if any of these are incorrect, for whatever reason, it is easy to go down the wrong design route, especially if this occurs early in the design process. By checking with your Certifier, such assumptions can be checked for validity.

Certification of a building gives all parties peace of mind that it will perform as predicted. This may add some value to the property as Passivhaus Certification becomes more widely recognised as a mark of quality. Certification of a single building in the UK costs from around £2,500, depending on the amount of Certifier input, which is worth clarifying at the outset.

Certification is more important in countries where the standard is not widely understood or

> ### Certification of buildings in the United States
>
> At the time of writing, the only US-based Passivhaus Certifier, the Passive House Institute United States (PHIUS), is not PHI-accredited.[1] US projects requiring PHI-accredited Passivhaus Certification now need to go to a certifier outside the United States. However, the PHIUS has formulated its own certification scheme, known as PHIUS+[2] – a Passivhaus certification that also includes the US-accredited rating system the **Home Energy Rating System (HERS)**. PHIUS hopes that by linking to the HERS it will increase the recognition of Passivhaus by association, and that someone building to the Passivhaus standard might thereby be eligible for any available US incentives and also be in a better position to qualify under the US **Leadership in Energy and Environmental Design (LEED)** rating system.

Passivhaus Certificate.

Certified Passivhaus Component logo.

recognised. In Austria, where the Passivhaus standard is very common, only one in five Passivhaus buildings now gets certified.

Certification of building components

Typically, the certification of building components[3] pertains to mechanical ventilation with heat recovery (MVHR) units, compact units, whole windows, window spacers and window frames. Manufacturers apply to the PHI to get certification. Component certification is priced to encourage manufacturers wanting to sell Passivhaus-certified products. For example, the PHI offers window certification from just over €2,500 (approximately £2,000).

Certified components often cost more than non-certified ones, so a judgement has to be made about the relative benefits and costs of using certified products. This is part of the role of the Passivhaus Designer. There is no requirement to use certified products (except in certain circumstances with the EnerPHit retrofit standard). However, it is worth noting that if a window manufacturer, for example, has taken the trouble to certify some of its products, it is more likely to be able to provide the technical information needed to calculate their energy impact in the PHPP. If selecting a non-certified window, you may be able to obtain an independently verfiable performance specification, but currently it is difficult to obtain this information for many products. Without independently verified figures, it is harder to reach the Passivhaus standard because much more conservative assumptions about the component's performance have to be entered into the PHPP. It is strongly recommended that you always use a certified model of the MVHR unit or compact unit. The detailed criteria for a Passivhaus-certified MVHR are set out in Chapter 7, page 106.

Certification of very low thermal bridge components

At the time of writing, very few products have been certified in this category.[4] As we will see in Chapter 8, when building a Passivhaus the aim is to avoid thermal bridges by careful design rather than to use expensive components. Certain types of construction, such as balconies and similar cantilevered designs, make thermal bridge elimination impractical, and in these situations there are certified products that can help. Alternatively, there are other solutions, although they may take the design down a different aesthetic route – for example, balconies can be supported independently of the main structure of the building, as shown in the photograph opposite.

Certification of Designers/ Consultants and contractors

There are two ways of becoming a Certified Passivhaus Designer or Consultant. The first involves a short but intensive period of study[5] followed by a demanding three-hour written exam. The courses (see Resources) are provided by PHI-accredited Certified European Passive House (CEPH) course providers and the exam is issued and graded by the PHI. The second way involves completing a Passivhaus project that achieves certified status, although this is a more difficult route. Certified Passivhaus Designers are usually architects or others who have specific proven building-focused qualifications, such as civil engineers or building technicians. Certified Passivhaus Consultants come from a wider variety of backgrounds, not necessarily related to construction. In either case, the content of the training and the examination is the same. The PHI's intention in using the two distinct professional titles is to convey to clients that Passivhaus Designers are able to design a Passivhaus (including doing the PHPP energy modelling – see Chapter 7) and Passivhaus Consultants are able to provide professional support to architects who are not trained in Passivhaus. An architect who is Passivhaus-qualified may also choose to 'outsource' the PHPP calculations to another Passivhaus Designer or Consultant.

Passivhaus Designers and Consultants have to renew their qualification by completing at least one Certified Passivhaus project every five years. They will be competent in using the PHPP, and will also add value to a project by providing consultancy to a design team without Passivhaus experience. Enlisting the help of a Certified Designer or Consultant can reduce the risk of the project failing to achieve the Passivhaus standard, and should save money on the build.

As of 2012, there is now a PHI-accredited Certified Passivhaus Tradesperson qualification (see Resources, PHI). There are a number of providers offering courses leading to the qualification, including a few in the UK and Ireland.

The certification process for Passivhaus buildings

The decision to certify a building can be taken after it is constructed. However, this approach is risky and may deny the project the added value from the input of the Passivhaus Certifier during design and construction.

Ideally, the decision about whether to get the building certified should be taken from day one (see Figure 3.1 on page 45). The pre-planning check shown in Figure 3.1 is usually at extra cost, but may be useful if the project is your first Passivhaus build, as it would provide expert input right at the start.

Example of structurally independent balconies (building in foreground) and thermally bridged balconies (building in background). *Image: WARM: Low Energy Building Practice*

Rather than reviewing and checking the Passivhaus Designer's completed PHPP, the Certifier normally produces his or her own separate PHPP calculation using information provided by the Passivhaus Designer, as an independent check. The box below shows the information that the Certifier needs in order to certify that the building meets the Passivhaus space heating requirement (annual specific space heat demand of 15kWh/m².a) in a typical single family house development.

Information required by the Certifier

- Block plan at 1:1250 and/or site plan at 1:200 showing due north and any shading objects.
- 1:100 elevations for each façade.
- 1:50 floor plans showing the internal layout for each floor.
- 1:50 sections showing the airtightness layer and airtightness sequencing.
- Plans showing MVHR ducting (exhaust, intake, extract and supply, including the location of sound attenuators).
- 1:10 and 1:5 plans of all junctions between opaque building elements, and between opaque elements and the doors and windows (head, jamb and sill details).
- Documentary evidence of the types of insulation used, including conductivity value certificates with CE marks (see Chapter 7, page 95) and invoices/delivery notes. Enough information for the Certifier to enter wall, floor and roof constructions (for the U-values worksheet of the PHPP).
- Photographic evidence of how and where insulation was installed and of how building junctions/thermal bridges were managed.
- Thermal bridge calculations and associated information, where there are junctions that are deemed not to be thermal-bridge-free.
- Information about the windows, including certificates for the key window properties (see Chapter 11) and a **window schedule** (a list of all windows in the building) showing window designations and dimensions.
- Information about the MVHR system (see Chapter 12), including an invoice/delivery note and a commissioning certificate.
- Airtightness certificate showing q_{50} (air permeability – see Chapter 9) for the completed building and a calculation of the internal volume.
- Signed declarations by the Passivhaus Designer / PHPP modeller and building site supervisor.
- If a specific climate dataset has been used for the project, the Certifier will need it, plus evidence of how it was produced (see Chapter 7, page 108).

In order to calculate the total annual primary energy demand, the Certifier also needs:

- Details of: the hot water heating system, properties of the hot water store and any solar thermal facility, and efficiency of the boiler or other heat source.
- Plans or a table showing the lengths of hot water pipes used in the building.
- Information about appliances and lighting.

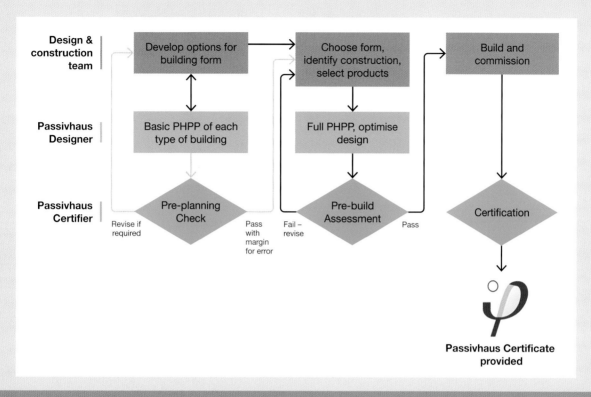

Figure 3.1 The certification process for a Passivhaus. Note the separate roles of the design & construction team, the Passivhaus Designer and the Passivhaus Certifier. *Source: WARM: Low Energy Building Practice*[6]

RECAP

In a country such as the UK, where Passivhaus is relatively new, certification is particularly beneficial. It provides a quality-control system as well as a framework for spreading low-energy-design skills. There are many claims made about buildings and their eco-credentials, and certification helps to protect against 'greenwash', which unfortunately is still common in a culture where we are not yet well informed about low-energy design. Certification may become advantageous when selling on a Passivhaus, as awareness of the concept spreads.

Certification in the UK can apply to whole buildings, building components and people. Designer or Consultant qualifications can be gained either through study or through completion of a Certified Passivhaus project. A building can be certified after completion, but there are considerable advantages to involving the Passivhaus Certifier from the beginning of the project.

CHAPTER FOUR
Challenges of meeting the Passivhaus standard

UK building culture, education and employment in construction, professional expertise, risk-aversity, the planning system, vernacular styles, the team approach, the disadvantages of a Passivhaus, Passivhaus and other low-energy standards

There are some notable challenges to address if deciding to adopt the Passivhaus standard. These are outlined in this chapter and, where appropriate, possible solutions are presented. Being aware of these issues is important, and you may need to consider early on what your approach to them will be. Some challenges are more cultural in nature and as such can only be managed rather than eliminated.

The UK building culture

House building in the UK has undergone a radical transformation from local building firms in the 1930s, through regional developers in the 1960s to national house building companies from the late 2000s. The housing sector is now dominated by large-scale housing developers, with only 50 companies building more than 500 homes each annually. This is different from the situation in most other European countries, where small-scale, more localised building firms remain the norm. The self-build, or self-procurement, model currently contributes to approximately 12 per cent of total new housing in the UK (although growing slowly), while this sector is much larger in European countries: a minimum of 50 per cent. (The concept of self-build extends from those who wish to physically build the house themselves to those who purchase a plot but appoint a team to carry out the construction work.) Austrian self-builders are responsible for 80 per cent of new houses! The US also has a strong self-build sector. In the UK, the number of those aspiring to the self-build model is significantly higher than the number achieving it – with only one in four managing to secure the necessary funding.

It would be helpful to Passivhaus in the UK if the self-build sector could be expanded significantly – to double or treble in size over the next few years. In our view, it is here that the greatest potential for good-quality, new low-energy houses lies. Until government legislation enforces such a standard, it is the self-builder or smaller builder who is most likely to have the interest and motivation to learn and adapt to a new innovative approach, and to invest the additional time and care needed to be successful in achieving the standard.

The current Conservative-led government's expressed commitment to localism and empowering local communities should represent an ideal opportunity to introduce new, creative ways to encourage these smaller initiatives. The government has recognised this and in 2011 committed to doubling the UK's self-build sector over the next ten years. This can only be achieved by councils and other large landowners introducing different mechanisms for releasing land for housing, thus providing alternatives to the normal profit-driven developer model. There are already highly successful schemes in Europe, where land for release is divided up into smaller plots (including single plots) for bidding and purchase by self-builders rather than being tendered out to single large national or regional builders. The largest self-build, local-authority-led scheme to date is in Almere (The Netherlands), a scheme currently with 800 new homes and thousands more due to follow.[1] There should also be greater opportunity for community schemes to build houses together in 'collectives' such as co-housing and Community Land Trusts, and much more pressure on councils and other landowners to deliver land to these groups.

At present, however, the most common pattern for delivery of houses in the UK is that house builders do not retain any long-term interest in the property they build. This means that the cost benefits of 'low energy in use' are not directly linked to the builder. It is apparent too that UK ownership patterns need to be revisited – with

> ### Self-build and 'custom-build' legislation
>
> In the latest draft National Planning Policy Framework (July 2011), the UK housing minister proposed that an obligation should be placed on local councils to meet the self-build demand – a potentially positive development. The Housing Strategy paper 'Laying the foundations' (published November 2011) further supports increased self-build, here sometimes referred to as custom-build, especially when exploring self-build for multiple units. Local authorities and other land-owning government bodies, such as the Homes and Communities Agency (HCA), can now release large plots of land for custom-build developments. The HCA has, as a result, included self-build as part of its land disposal strategy. A few councils are now trialling schemes releasing self-build plots and assisted self-build, Swindon council being the first. These transactions could easily be tied into a requirement to build to an ultra-low-energy standard. We hope that these early model schemes will lead to wider adoption of such an approach.

high house costs and high economic debt (both personal and governmental), there is an urgent need for a radical rethinking of our patterns of housing supply and ownership.

Education and the construction workforce

We do not significantly invest in education and training of the construction labour force in the UK, and this sadly reflects a lack of recognition of the skill and value of the 'practical', hands-on professions. In the UK, our love affair with intellectual knowledge and university education has devalued many with enormous intelligence and ability who are vital to our economic resilience. The separation of practical and theoretical knowledge becomes increasingly problematic once higher performance standards are required. Currently we rely on much of the workforce carrying out their work in isolation from other skill sets, but in order to successfully achieve low-energy standards, it is essential that the whole workforce is included in an understanding of the principles of low-energy building. Without this it is highly unlikely that individuals will significantly alter their normal working practices. The need for coordination and understanding of the interfaces between packages of work becomes much more critical on a low-energy build, and is particularly relevant if attempting to achieve demanding airtightness levels. This reality is being borne out as the Passivhaus standard is applied to larger UK projects, where multiple housing units are involved and contractors are struggling to achieve the airtightness levels needed.

In some European countries, such as Germany, there are training schools for each construction trade and these are compulsory for builders. This type of educational network provides an effective framework for delivering appropriate training and for widely disseminating knowledge of best building practice. Unfortunately, in the UK we do not have such extensive training networks in place. The National House Building Council (NHBC) was founded before the Second World War to address poor building standards and now provides a warranty system, but this reflects current Building Regulations only. A national organisation such as the NHBC could again become a network for delivering the more thorough training required for very low-energy building. It is possible for the State to introduce such training programmes, as evidenced by the Canadian government, which offered training on low-energy housing delivery as far back as the 1980s. In ten years, 70 per cent of Canadian building contractors attended a voluntary R2000/Super E training course – a notable success.

The Passivhaus Institut now offers a Passivhaus qualification for tradespeople: the first three-day PHI training course (and exam) took place in January 2012 in Germany. The course is oriented to specific trade skills for Passivhaus – mainly thermal insulation, window installation and heat supply/ventilation. There are a few certified providers of the course in the UK and Ireland.

In the UK, the **AECB** (the Sustainable Building Association) set up a series of modular Passivhaus training courses in 2011 through its CarbonLite Programme (see Resources, AECB), one of which, called 'Construction of Passivhaus', is for contractors. Some suppliers of low-energy products, such as airtightness tapes and membranes, will also offer on-site 'toolbox' talks, often at no extra cost. These types of initiative are essential to ensuring the successful delivery of low-energy projects. However, to reach anywhere near the level of success that the Canadian initiative achieved requires broader national policy and support.

Changes in the employment structures of workers

Economic pressures have radically reduced the numbers of construction workers (plasterers, carpenters, electricians, etc.) directly employed by building firms. Many work on day rates and rely on moving from site to site as their services are needed. The transitory nature of the workforce means that working teams do not remain static but are constantly in flux. In these circumstances, controlling day-to-day activity on-site is more difficult, as is ensuring that all are fully informed of project particulars. Individual workers are less likely to feel committed to larger construction companies when there is no long-term job security or continuity, and in these circumstances engendering team spirit becomes a greater challenge. In particular, the attention to detail required to achieve Passivhaus airtightness standards is made more difficult on-site, as each trade must be involved in protecting earlier work and avoiding subsequent damage, which requires very good levels of interaction and communication. The damage–repair cycle, a pattern very common on UK building sites, is fatal for an ultra-low-energy build.

The professional design team

Some comment on the professional design team must also be made here. The construction details proposed can sometimes lack practicality on-site – how buildable are the solutions proposed? When it comes to low-energy design, it can be argued that making traditional detailing solutions more complex can quickly become more problematic than rethinking from basic principles and proposing alternative methods of building that are then simpler and easier to replicate on-site.

Designers, architects, engineers, services consultants and quantity surveyors can simply lack knowledge of low-energy design, and be unwilling to admit a limited understanding. Training needs to be undertaken by both the professional and manual trades. Rethinking our approach to design means accepting that what we have been doing no longer meets the required standard. Low-energy design is not a one-off luxury but a step change in the quality of build.

It may also be that the normal clause that allows contractors to use specified materials or 'equal and approved' ones has to be reconsidered. This has led to the common practice of sourcing alternative products on-site, often without reference back to the design team, which introduces greater risks for a low-energy building, where the performance of particular products is essential.

Risk-averse attitudes

House builders work in a difficult sector, being directly affected by house-price volatility while also having to manage various site-specific risks (land acquisition, gaining planning permission and unpredictable construction costs). For these reasons there is an understandable and general reluctance to innovate. The financial sector considers house building a high-risk activity and therefore it could be argued that a conservative, risk-averse building industry is justified. At the same time, the need for innovation is widely recognised – especially if low-energy housing is to be widely delivered. When new, unfamiliar construction methods and systems are being introduced, this creates another level of risk (both of potential failures and cost escalation) and without government pressure there will be a natural resistance to change.

The planning system

The UK planning system currently tends to inhibit the adoption of innovative design solutions. Some of the reasons for this are as follows.

- Processing of applications is slow, despite government targets, and this has a direct economic impact – in a recent review of housing case studies commissioned by the Royal Institution of Chartered Surveyors (RICS) in December 2010,[2] more than half encountered substantial problems when trying to process planning applications. This tends to encourage applicants to submit tried-and-tested design solutions that can aid a smooth ride through the planning process. Investing and designing carries risk.
- Planning authorities encounter real problems in recruiting knowledgeable planners. The job is stressful and might be considered unrewarding (planners get a lot of abuse!). As yet there are few planners who are knowledgeable about how low-energy performance might impact on design.
- The guidelines and general planning approach currently lead to a tick-box mentality (although this may now be changing under the government's 'localism' agenda). There is a general presumption that what already exists is best (blend in with it), including a strong conservation bias characterised by traditional aesthetic sensibilities.

Of course, there are government planning policies that reflect the sustainability agenda, but as yet there is uncertainty as to what weight these will or should carry, especially when balanced against other planning considerations. The impact of these newer policies currently remains stifled by much more established planning policies, especially those involving aesthetics and conservation. Unfortunately, ultra-low-energy design is not a 'bolt-on' extra and as such does have aesthetic implications – in particular, deeper eave overhangs and window shading devices. The entire orientation of the building becomes much more significant, which can affect both window locations and sizes. There has to be a new consideration of what the balance should be between cultural preferences for 'blending in' with traditional styles and accepting the consequences of opting for increased energy performance. For planners to begin to make these judgements intelligently requires clearer government policy and guidance,

Chapter Four • Challenges of meeting the Passivhaus standard 51

along with a better understanding of the design principles and potential visual effects.

The planning system could play a key role in encouraging more innovative low-energy housing applications (this is further discussed in Chapter 14), although, with the current movement towards deregulation of the system, this would be more difficult and the system could become even more unpredictable. With the localism agenda, the onus will increasingly be on community interest groups to lobby and persuade the wider community of the benefits of innovative or culturally challenging projects – this process may well require increased time, energy and money from applicants. This then favours those with more power and those who can better afford financial investment with the risk of refusal. Without clear policy and regulation, low-energy design is in danger of continuing to be viewed as an optional extra and as the interest of a minority group rather than an accepted standard. Whether the current UK government continues with Labour's plan to adopt the zero carbon standard in 2016 will be pivotal to this.

UK vernacular styles

There are strong vernacular housing styles in the UK that vary region by region, often reflecting local materials (e.g. brick or stone) and particular climatic conditions. These can sit very deeply in our collective unconscious and contribute to a general feeling of security and well-being, creating a sense of place and 'home'. Venturing into new, unfamiliar architectural styles can easily (perhaps inevitably) create waves of unsettled feelings. However, simply directing our energy into maintaining the past is not a long-term viable strategy – evolution (and cultural evolution) itself shows us the inevitability of change.

But we do not need to throw out the baby with the bathwater. There is much to be done in combining local materials and styles with new low-energy principles. Passivhaus is not a global design aesthetic, but it will bring its influence to bear on the finished design. A good designer will do this with sensitivity and not bludgeon local 'flavours' to death. (Although there must also be a place for some radical contemporary designs if the context is appropriate.) A perception of the Passivhaus approach could be that its scientific roots have not always been balanced with aesthetic finesse, and this may create an unfortunate impression that the two cannot be wedded. However, problems of poor design are not limited to Passivhaus! In the next few years we hope to see more good examples of well-applied, ultra-low-energy principles coupled with excellent design and intermingled with local or regional flavour. The images overleaf show a variety of Passivhaus projects that demonstrate the design diversity already being realised in the UK.

> **Ultra-low-energy design is not a 'bolt-on' extra, and as such does have aesthetic implications**

Top left: The Larch House, Ebbw Vale – Welsh social house scheme using a palette of local materials. *Image: Jefferson Smith* **Top right:** Y Foel – detached house set in a secluded site in Wales. The first Certified Passivhaus in the UK. *Image: JPW Construction* **Bottom:** Underhill – contemporary solution addressing the visual constraints of the site in an AONB. *Image: Samuel Ashfield*

Top left: Grove Cottage – the first Certified EnerPHit-standard retrofit of a solid-walled Victorian (1869) town house. *Image: Simmonds.Mills*   **Top right:** Camden Passivhaus – Contemporary design. The first Certified Passivhaus in London. *Image: Jefferson Smith*   **Bottom:** Totnes Passivhaus – Adam Dadeby and Erica Aslett's home. Retrofit reflecting existing styles on a 1970s estate attributed with architectural merit. *Image: Malcolm Baldwin*

There will always be a place for some historic examples – those with particular architectural merit – to be preserved and maintained. These houses perhaps need to be lived in and upgraded in a manner that is in keeping with their original construction and design. Applying the Passivhaus standard to all buildings is not appropriate or realistic.

Bringing the right team together

If your intention is to build an ultra-low-energy home, then one of your first tasks must be to bring the right team together. These key early decisions will either set the project off on a useful trajectory or send you on a road to inevitable frustration, whereby failure can be a very real possibility. Turning round and retracing your steps becomes extremely problematic, if not impossible, as time marches on.

Make sure that someone in your design team is properly trained in Passivhaus principles and methodology. Any others who have no experience of low-energy design need to be committed to the concept (not just giving a nod to please you) and keen to put in the extra time to learn and reskill for the project. If they are, then it will be an enjoyable, if demanding, journey. Set the project up as a low-energy build from day one and spend enough time on the initial planning, on setting up the PHPP tool and working out your air leakage strategy. The more time spent here, the fewer mistakes or problems there will be on-site (these will cost you far more than extra design time). Taking all this into account, in our view you should not need to pay much more than for an equivalent conventional build (see Chapter 2). This is where tendering can be a double-edged sword – a low price seems great at the time, but when you are aiming to create a quality build such as a Passivhaus, people need to spend the time to achieve it. Corner-cutting will almost certainly create havoc with your initial aims. Bringing the right team together is an extremely important part of the project; this is covered in more detail in Chapter 6.

The disadvantages of a Passivhaus

Yes, there are some disadvantages to a Passivhaus build! These are discussed here.

Inflexibility of the ventilation system

The MVHR ventilation system is not straightforward to adapt, so if subsequently you wanted to alter the internal layout of your building you would need to invest in modifying your system and then arrange for the system to be rebalanced (some newer MVHR systems are easier to rebalance and to modify the ducts). Once the ventilation **ducts** (the pipes that run between the MVHR unit, the living areas and the supply/extract points) are built in, accessing them means affecting decorative finishes and making good afterwards. Of course, this is equally true for changes to plumbing and other building services.

Wanting a cooler bedroom

Some people like to sleep in cooler spaces than their general living areas, whereas Passivhaus is built on the principle of a single heating zone, with surface temperatures kept even throughout the building. The solution adopted by most is simply to open the bedroom window at night. This will carry a very small energy penalty, perhaps around 2kWh/m^2.a, but it will effectively reduce the temperature of the room so long as the external ambient temperature is lower than the internal.

If the space heating is being provided through the MVHR system (i.e. in the air supply), you could reduce the airflow to the bedroom in order to reduce the heat output (and fulfil the ventilation requirement by opening windows), or bypass the heating element altogether – although this would be a permanent and therefore inflexible arrangement. It will be simpler to bypass the heating element with some MVHR units than with others.

Imported components

At the time of writing, certain Passivhaus essential components are available only from companies outside the UK (and the USA), and some people will feel uncomfortable about the additional energy involved in shipping these items. All the construction materials are available in the UK, but Passivhaus-certified windows, doors and MVHR units still have to be imported in nearly all cases. However, the situation is rapidly changing and there are already companies in Ireland and Wales that are setting up to supply and manufacture these Passivhaus-certified products. (For useful links to information on products and developments, visit our related website, www.passivhaushandbook.com.) If you want to source all products from the UK and Ireland, it may be some time before there is a huge choice, but we would expect that within the next couple of years it will be possible to source all the relevant components from within a narrower geographic radius.

Hand-built from 100-per-cent locally sourced materials?

There are some who will want to commit to building their home from locally sourced, natural materials, since this both supports local economies and keeps embodied energy levels as low as possible. ('Natural materials' normally refers to materials and products that have undergone minimal processing or are in their raw state. For more on natural and low-embodied-energy materials, see Chapter 5.) Some also want to build their home with their own hands and minimise the 'professional' input of contractors and architects – perhaps motivated by a need to keep costs to an absolute minimum and/or to gain the satisfaction that such an enterprise can bring to the self-builder. There are no issues with using materials sourced locally or using natural materials for new houses, but, as noted already, there will be some elements (windows, MVHR) that will almost certainly need to be sourced from a wider geographical area. Passivhaus will not cope with low-performing windows, for example, or without the MVHR – these elements work together in an integrated approach. Furthermore, if you are carrying out a retrofit scheme, the existing fabric may necessitate using higher-embodied-energy materials to meet the thermal performance requirements. This may be for structural reasons or due to height or depth restrictions.

While professional workforces can produce surprisingly shoddy work(!), it is also true that the practical trades all carry learned skills, which are not simply picked up (unless you really are talented). Putting together an airtight building with no thermal bridges requires design-detailing skills plus the skill to execute the construction in the correct sequence and with some accuracy on-site. It is true that many self-builders pay greater attention to detail than paid construction workers, mainly because of their personal involvement and high level of commitment to their project, and Passivhaus could be well suited to such a person. However, the support of some skilled workers is, in our view, necessary for a successful Passivhaus build. A self-builder with some professional support may even be the ideal Passivhaus builder, given the realities of the current UK building culture.

However, if you want to build your home at very low cost and with minimal or no skilled support, then Passivhaus is probably not the right choice. That is not to say that it cannot be done, but such builds will always be niche and not mainstream. The concept behind Passivhaus is that it is capable of being adopted as a common mainstream standard.

The invasive and extreme nature of retrofit

If you undertake a retrofit project, it will involve a very invasive stripping-back of your building – taking down ceilings, etc., and getting down to the bare structure of the house – in order to meet the demanding Passivhaus or EnerPHit standards. The challenges of removing and minimising thermal bridges and ensuring continuity of the insulation and airtightness barrier make the project much more demanding than a superficial refurbishment would be. This inevitably pushes up costs, and with retrofit it is also difficult to accurately assess all the necessary work prior to opening up the building. The task, therefore, is not for the faint-hearted. You are aiming for a revitalised, building with a hundred-year lifespan, not a facelift lasting five to ten years.

In the case of retrofit, and especially for one-off projects, the rigours of the Passivhaus standard need to be considered in terms of economic viability. Councils and registered social landlords (RSLs) who can tackle groups of existing housing are in a better position to adopt the standard. Government incentives (particularly VAT parity with new-build for retrofit) would assist, as would changes to the planning system (see Chapter 14). However, it is possible to meet the target – for example, a single semi-detached retrofit in Ealing, London, has achieved 0.44ach in its final air test in May 2012 at favourable costs, but with a huge personal input of time and energy by the owners. With VAT set at 20 per cent for retrofits, there becomes an undue bias towards demolition, if possible, since new-build is zero-rated for VAT.

Problems in comparing Passivhaus with other low-energy standards

'Low energy' and 'eco' have become generalised terms that can refer to many types of buildings and differing philosophical outlooks. An eco-house may simply mean that a greater proportion of natural materials have been used or that a renewable energy source, e.g. a solar hot water system, has been incorporated into the design. There are no common agreed energy-performance standards for an eco-build and therefore no means of making quantifiable comparisons with the Passivhaus standard.

In the UK, the term 'low energy' is most commonly linked to and defined by the Code for Sustainable Homes (CSH). The CSH does contain parameters for energy performance, using the **Standard Assessment Procedure (SAP)**. SAP calculations have been required for UK Building Regulations Part L1A since 1995, and the most recent version of SAP came into force in October 2010. (Building Regulations Part L1A is the standards relating to energy/carbon emissions for new dwellings). SAP is used to predict the energy used for heating, hot water and lighting, and is given as a number between 1 and 120. The rating achieved is affected by a variety of inputs (the thermal efficiency of the envelope, the level of airtightness, the choice of fuel, the efficiency rating of the hot water and heating system, and any renewable energy sources).

SAP is also used to check compliance with other more specific requirements as set out in the Building Regulations, such as minimum U-values (see Chapter 11), but most significantly it is used to predict carbon dioxide emissions. SAP results

Chapter Four • Challenges of meeting the Passivhaus standard 57

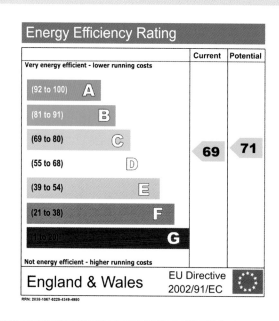
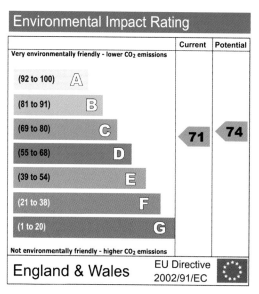

A to G banding for energy performance of buildings, using the Standard Assessment Procedure (SAP).

are translated into two banding or rating systems from A to G, with band A having a rating of 92+. The Environmental Impact Rating reflects the predicted carbon emissions, while the Energy Efficiency Rating translates the results into predicted running costs. The Building Regulations set the allowable carbon emission rates. A Target Emission Rate (TER) is compared with a Dwelling Emission Rate (DER). The TER is based on the emissions of a notional dwelling built to 2006 Building Regulations, less approximately 25 per cent (depending on your fuel choice). The 'less 25 per cent' comes from the 2010 Building Regulations, where the government wanted compliant buildings to be 25-per-cent better than 2006 compliant dwellings. The TER and DER are measured in CO_2 kg/m².a (kilograms of CO_2 per square metre [of internal floor space] per annum). From this brief account it should be clear that it is a rather convoluted assessment concept!

Using SAP as the tool, the UK has so far chosen to assess energy efficiency in terms of carbon dioxide emissions. Measuring CO_2 emissions, as set out within SAP, further involves inputting a wide range of data, some of which requires assumptions to be made by the assessor. The fact that the carbon emission levels are influenced by your fuel choice demonstrates that the resultant DER is not a direct calculation of building fabric performance. SAP and PHPP also have different measuring rules, e.g. SAP fabric measurements for external walls use internal dimensions, while in the PHPP external dimensions are used. The treated floor areas (TFA) are also measured differently – more favourably in SAP (see Chapter 7, page 94). SAP currently requires no site-specific climate data (likely to change soon), unlike the PHPP, and climate has a profound impact on energy consumption, even within the UK.

SAP fulfils the European requirement that each member country has a single energy assessment procedure – it is essentially a compliance tool set against a notional standard (a 2006 compliant dwelling). It is not a tool for designing buildings and should be used as such only in very broad terms. It was designed for use with conventional buildings and does not demand accurate input of detail in the same way that the PHPP does – a SAP calculation on a typical modest house might take two hours, compared with a minimum of two days for the PHPP. The latter was specifically formulated for ultra-low-energy buildings and to set building performance (regardless of fuel type or use of renewable energy sources) and is both flexible and accurate. PHPP is both a design and compliance tool; SAP is not as sophisticated.

There is a strong sense among many who are involved with low-energy buildings that measurements should move away from carbon-based to energy-based measurements. The primary reason for this is that a space heating target measured in kWh/m².a, such as is required for Passivhaus, directly drives construction towards better fabric performance: minimal air leakage, minimum thermal bridging and optimised contiguous (continuous) insulation of the thermal envelope. The Passivhaus methodology of using the PHPP tool to measure energy for space heating is therefore very useful to any low-energy design – up to a performance level of approximately 40kWh/m².a (assumptions made in the modelling software may not be valid above this level). You may, for example, be aiming to build a Level 5 or Level 6 home as set out in the CSH, or be using another energy/sustainability assessment system. The Passivhaus methodology will ensure that the focus is on the building fabric, not on other matters such as bolt-on renewable-energy inputs or using low-carbon-emitting fuels. For this reason, the Zero Carbon Hub (see page 66) has advised the UK government to adopt a Fabric Energy Efficiency Standard (FEES), measured in kWh/m².a, as part of the CSH standard and for Building Regulations from 2013.

The Passivhaus low-energy target is not just about reducing emissions, it's about construction quality

It is important to understand that the purpose of the Passivhaus energy target is wider than simply reducing carbon emissions. The low-energy target drives the quality of the construction, which in turn contributes to a good interior climate and helps to eliminate potential moisture and overheating risks. Getting the fabric right is not an optional extra in a low-energy build. Even if not using the PHPP, applying basic Passivhaus techniques – designing an efficient building form, removing thermal bridges and reducing air leakage – will be valuable. We would still suggest that you include PHPP modelling as a design tool if you are intending to build an airtight home and use whole-house heat recovery ventilation (MVHR) – even if your preferred sustainable assessment system is not Passivhaus per se (see Chapter 7, page 91).

Chapter Four • Challenges of meeting the Passivhaus standard

RECAP

Is Passivhaus always the right choice? It certainly cannot always be the right choice, and if you do decide to adopt the standard there are a variety of challenges to be faced, including working within the prevailing building culture and the planning system, and ensuring you bring together an appropriate, committed and knowledgeable team. There are potential disadvantages of a Passivhaus build, from the invasive nature of the changes that need to be made in a retrofit project to the inflexibility of the ventilation system, but generally these can be acceptably addressed.

In terms of architectural detailing, Passivhaus is an energy standard, not a design style. Balancing function with beauty is the domain of the architect and is always a challenge, whatever the project. Adapting and integrating local vernacular styles is part of that aim.

Perhaps the biggest challenge – one that goes far beyond the specifics of a Passivhaus build but is clearly highlighted by a consideration of Passivhaus – is in finding new ways to deliver our future UK housing stock. This must include the way land and funding is made available and must broaden the housing sector out from the few large regional and national house builders.

If governments choose to adopt alternative low-energy standards, it is important that the principles of Passivhaus are understood as essential to any low-energy approach. These principles can then be intelligently incorporated within those alternative standards.

CHAPTER FIVE
Natural materials, zero carbon and resilience

Natural and low-embodied-energy building materials, zero carbon and the Code for Sustainable Homes (CSH), on-site low- or zero-carbon energy, post-peak energy, energy returned on energy invested (EROEI)

While Passivhaus does not address the question of the type of materials used and the energy embodied in those materials, this is an area of interest for many who might also be interested in building a Passivhaus. Concern about using materials with low embodied energy, and more locally sourced materials, relates to the wider issue of the resilience of our preferred solutions to interruptions in resource supplies. The UK zero carbon standard suggests a very low-impact housing solution, and the question of how this relates to Passivhaus is therefore pertinent.

Zero carbon is the chosen UK government's strategy for achieving low-energy housing in the future, and is planned to become a statutory requirement from 2016. For anyone interested in building a Passivhaus in the UK, understanding how the zero carbon and Passivhaus standards compare and might relate to one another will therefore soon be a necessity. A step change towards zero carbon is planned for 2013, so this need is already imminent.

Building materials

Passivhaus does not stipulate the building materials you should use in your construction. The standard is focused on 'energy in use' because, in standard builds, this is by far the largest proportion of the energy used by our houses over their lifetimes. It is also an area of energy consumption we have historically failed to reduce effectively – in the UK, even eco-housing has tended to consume far more energy for space heating than was the original design intent. Also, as mentioned in Chapter 1, one of the goals of the Passivhaus Institut is to get the Passivhaus standard adopted as widely as possible, which means engaging with the mainstream construction sector. If the Passivhaus standard also stipulated that only low-embodied-energy and low-impact building materials were to be used, it would strongly discourage mainstream builders and risk making Passivhaus an eco-niche. However, there is no reason why a Passivhaus must be built with high-tech, high-impact, 'non-natural' materials.

Embodied energy

The embodied energy of a material is the sum of all the energy inputs required for its production – from extraction to fabrication and transport, etc. The more highly processed a product, the higher its embodied energy is likely to be. A material such as steel, for example, will have a very high embodied energy (extraction, heated to high temperatures, heavy to transport), while a locally sourced straw bale will have a very low embodied energy (minimal processing, lightweight). Appropriate use of high-embodied-energy products and materials should be a consideration for anyone involved in building, and it seems only responsible to use low-embodied-energy materials and products wherever possible.

There are of course 'added' values that a high-embodied-energy product might bring, including improved performance, structural strength, space saving or aesthetic benefits (the latter being less easy to quantify). Natural materials can also bring added value, especially in the way they can modulate air quality and handle moisture levels (see Chapter 10). Your choice of materials and products should, ideally, involve a sensible consideration of all these factors. In practice, financial and planning constraints will also often influence decisions.

Measuring embodied energy can be quite complex and the methodologies differ – how far do you go in including every input involved? There are 'cradle-to-grave' calculations (full life-cycle

Practical ways to reduce embodied energy

Clearly, in an ultra-low-energy build there is particular value in reducing or minimising the energy used in construction, provided it does not compromise the building's energy-in-use performance, as to do so would quickly become counter-productive. In addition to using low-embodied-energy materials, there are also other ways to reduce embodied energy, if this is of concern to you.

- Avoid over-ordering materials.
- Reuse or recycle materials wherever possible.
- Retrofit and extend rather than demolish (if the existing structure warrants it).
- Use locally sourced materials, where possible.
- Use less highly processed materials, where possible.
- Avoid over-engineered solutions – especially steelwork and concrete foundations.
- Use assemblies that can be dismantled rather than destroyed (bricks with lime mortar is a good example of this, as the lime allows the bricks to be reused).
- Use 'Lifetime Homes' (a standard that aims to ensure that homes are suitable for the whole lifespan of occupants)[3] design principles for creating flexible and adaptable spaces.
- Consider how big your house really needs to be.
- Build once – build well! Quality is often a good long-term energy-saving strategy.

analysis), 'cradle-to-site' and 'cradle-to-gate' (up to the point when it leaves the factory). But whatever the methodology, the principle is clear and the data useful – within the UK, the CSH, **BREEAM** EcoHomes and the US LEED assessment systems all rate materials according to their embodied energy.

The University of Bath (UK) has produced an 'Inventory of Carbon and Energy' database for hundreds of materials, based on energy inputs into their manufacture, together with the energy costs for transport.[1] The figures are often given as ranges because so many inputs are variable that single assessments are not feasible.

As already noted, the embodied energy of the materials used to construct a house is far less than the energy that will be used in the house over its lifetime. Based on a three-bedroom standard house, BRE estimated in 1991 that energy in use would exceed embodied energy by 12 to 30 times, assuming a 60-year life.[2] Based on this assessment, energy in use will overtake embodied energy in a maximum period of 5 years. Other studies may differ on the exact figures, but all clearly demonstrate that minimising energy consumption in use is much more critical than reducing embodied energy. It follows that there is room for some pragmatism in the choice of products, where higher-embodied-energy products can help to reduce energy in use.

Once a house is built to ultra-low-energy standards (a space-heating target of less than 40kWh/m².a), the proportion of energy used represented by embodied energy becomes more significant. For a 'typical' house the embodied energy is up to 10 per cent of total energy over its lifetime; for a low-energy house this may rise to 30 or 40 per cent.

When assessing embodied energy it can be helpful to consider whole building assemblies – combinations of building materials that make up a component such as a wall or roof construction – rather than just individual materials. Table 5.1 shows how some common building assemblies compare in terms of the embodied energy they represent. While these are not low-energy constructions, the equivalent low-energy assemblies would reflect similar differences. It is useful to note that heavyweight constructions tend to have higher embodied energy than lightweight constructions.

Use of high-embodied-energy components within buildings

The embodied energy of a material sometimes needs to be assessed in terms of the lifetime energy benefit it can bring. This might be in terms of energy it will generate (e.g. solar hot water panels) or energy it will save (e.g. insulation). Manufacturers of high-embodied-energy insulations will argue strongly that the enhanced performance (lower thermal conductivities than natural insulations) and the energy saved through the lifetime of the building make quibbling over the embodied energy of the product irrelevant. Some argue (cynically or realistically, according to your taste) that the fossil fuel energy will be consumed anyway, so it's better to use it to save energy in the future.

In retrofits, spatial and structural constraints imposed by the existing building structure can make the use of high-embodied-energy but high-performance insulation a pragmatic compromise, since this may allow the existing structure to be saved while still achieving an acceptable energy performance level.

Windows have high embodied energy compared

Table 5.1 Embodied energy in some common building assemblies

Assembly	Embodied energy	
	MJ/m²*	kWh/m²*
Clay brick wall with cavity	860	239
Concrete block wall with cavity	465	129
Timber-frame wall with timber rain screen	377	105
110mm concrete slab on ground	645	179
Suspended timber floor	293	81

* m² refers to the area of the building assembly.
Source: Lawson, B. (1996) *Building Materials, Energy and the Environment*

Terms explained

kWh/m² – kilowatt hours per square metre.
MJ/m² – megajoules per square metre. A megajoule (MJ) is a unit of energy. 3.6MJ are equal to 1kWh.

with many other building components. Clearly, though, a simple, single-glazed, wooden-framed window has a considerably lower embodied energy than a triple-glazed, argon-filled window with insulated wooden-aluminium frames. If embodied energy were the overriding criterion in our choice of components, we would be installing single-glazed or perhaps air-filled, double-glazed, wooden-framed windows, despite their poor energy-in-use performance.

Natural materials

What, in fact, do we mean when we talk about natural materials? Normally this refers to materials and products that have undergone minimal processing or are in their raw state; in addition, there may be greater opportunity to source such materials locally. Both these factors correspond to a significantly lower embodied energy.

Another attraction of using more natural materials is that they can contribute to better indoor air quality (IAQ) . Natural materials and products will usually not 'off-gas' in the way some products do (examples include various carpets, paints, stains, insulations, glues, etc.). **Off-gassing** is the evaporation of volatile chemicals at normal atmospheric pressure: a process that introduces contaminants into the indoor air. The air in a typical home contains a large range of these volatile organic compounds (VOCs). The more airtight the building, the greater the impact of off-gassing on air quality. Some natural materials can even improve air quality by absorbing VOCs and also by absorbing and de-absorbing (releasing) moisture, thereby helping to regulate internal humidity levels. We discuss these potential health and air-quality benefits in more detail in Chapters 9, 10 and 12. The health impacts of poor IAQ are widely acknowledged, and this is therefore a very important consideration when building a home.

Some natural building materials will be better suited to a Passivhaus build than others. We have not yet come across a straw-bale Passivhaus in the UK, for instance, although there are now examples of these with low airtightness levels. We understand that there are now Passivhaus-level straw-bale buildings in Austria (individual bale humidity and density measurements needed to be taken to ensure performance for certification). Part of the difficulty lies in managing the levels of movement common with certain natural materials, and ensuring that this does not compromise the airtightness strategy. Natural materials also tend to have lower thermal performance levels (the **lambda value** – a measure of **thermal conductivity** – of phenolic foam, say, is 0.021-0.024**W/mK** (watts per metre per degree kelvin) versus sheep's wool at 0.035-0.04W/mK – see Appendix B). Natural materials also have less consistent lambda values, which means that more conservative figures have to be used in the PHPP. All this makes a significant difference to the depth of insulation required to meet ultra-low-energy building targets.

Essentially, there is no reason why a Passivhaus cannot be built using natural materials for the majority of the construction elements. Timber-frame construction is commonly used in both Europe and the United States, and frames can be insulated with natural materials such as hemp, sheep's wool, woodfibre board or even recycled waste paper (cellulose). For Passivhaus, some elements of your building will necessarily need to be more highly processed, i.e. less 'natural'; in particular the triple-glazed windows (although these often have timber frames; it is also likely that they will be produced more locally in the future – see Chapter 4, page 55). Also, airtightness tapes use very specific glues that cannot be classed as 'natural'. Of course, even a 'natural' building is never 100-per-cent natural – there are always light fittings, kitchen appliances and heating systems, etc. to consider, all of which are manufactured using more processed materials.

Carbon sequestration by natural materials

Some natural materials sequester carbon, i.e. they store carbon within their structures as they grow (timber being an obvious example). If then used in a building, they effectively 'lock up' this carbon over the building's lifetime. Some materials are particularly good at sequestering carbon – hemp and straw being two notable examples, since they can be grown as crops on farms and have useful insulating qualities as well. Using such materials means that we can begin to see buildings as potential carbon stores. This is not always a straightforward equation, as the environmental benefit depends on whether the overall volume of the resource material is increasing as a result of your use or not – for example, increased demand for tropical hardwood is likely to have the opposite effect, i.e. of diminishing the area of rainforest. On the other hand, European-sourced timber, where the forest area is being managed and increased to match demand, will be effective. Building with timber is also a better use for our limited timber resource than burning it for fuel[4] – burning, of course, releases the carbon back into the atmosphere. If farmers could grow break crops (secondary crops, such as hemp, grown as part of a crop rotation system and often used as soil improvers) for use in buildings, this could help to increase the amount of carbon sequestered in our buildings.

Passivhaus and zero carbon

'Zero carbon' and 'Passivhaus' have both become popular marketing buzzwords in recent years, and many will be interested to know how the two relate to each other, which we explore here.

UK policy – the Code for Sustainable Homes (CSH)

Most relevant public policy and discourse has focused on reducing carbon dioxide emissions, in an attempt to respond to the global climate predicament. The UK government has legislated to make mandatory reductions in the UK's CO_2 emissions: to at least 80 per cent below 1990 levels by 2050, with an intermediate target of a 34-per-cent reduction by 2020. Reduction in CO_2 emissions is measured relative to 1990 levels.

Housing contributes approximately 30 per cent of the UK's total CO_2 emissions, of which about half is from space heating. The Code for Sustainable Homes (CSH) is the UK government's chosen vehicle to set reductions in CO_2 emissions from the housing sector, although this also incorporates broader sustainability measures. The CSH uses a weighted rating system, consisting of nine categories, which address anything from providing bike storage to installing a solar hot water system. Category One, 'Energy and Carbon Dioxide Emissions', is the most important of the nine.

Dwellings can be rated at one of six levels within the CSH system, Level 6 (L6) being the most stringent and termed 'zero carbon'. Zero carbon is currently set to come into force as a standard for new residential homes in 2016, with a similar requirement for non-residential new buildings to be in place by 2019.

The 'zero carbon' definition

The definition of zero carbon has been under review and discussion since the initial proposal of the standard. The intention has always been that it would mean 100-per-cent reduction in net emissions relative to a dwelling compliant with the 2006 Building Regulations (England and Wales) – according to the initial definition, this was where any emissions created were offset by those 'saved' using on-site renewable capacity. Under what exact terms this will eventually be enforced and what the uplift in cost might be (and to whom) remains unresolved at this time.

Carbon emissions have been separated into what are now termed **regulated** and **unregulated carbon emissions**. Regulated emissions are those from fixed building services, i.e. heating, ventilation and lighting; unregulated emissions are those relating to energy used by the building occupants, e.g. from cooking or electrical appliances. The initial plan was that 100-per-cent reduction in both would be the target – the L6 standard. Concerns about whether this was a realistic target in practice led the government in 2011 to alter the definition of zero carbon to include only the regulated carbon emissions. This means that zero carbon can now refer to either Level 5 or Level 6 of the CSH.

The Zero Carbon Hub (see Resources), a public–private partnership, has been working since 2008 to support the delivery of zero-carbon homes, and this includes the development of a final definition for zero carbon. The Hub's extensive work has included publishing various advisory papers and carrying out useful consultations.

Achieving 100-per-cent reduction in carbon emissions, even if from regulated emissions only, involves the significant use of on-site renewable energy sources. The practicalities of having enough roof space, not to mention the cost burden, has led to a further strategy being introduced – allowing a certain proportion of the emissions reductions to derive from off-site sources or what are termed 'allowable solutions'. What exactly these will be or how they will be delivered is again unknown, but will most likely involve a payment towards the cost of introducing carbon-saving projects. The proportion in carbon reductions that will have to be achieved

Zero carbon and stamp duty

The original 2006 zero carbon definition – relating to both regulated and unregulated emissions – is the one used by the UK's tax authorities (HMRC) to validate exemption from property sales tax (stamp duty) on homes below £500,000 in value.

by the house itself is termed 'carbon compliance'. The split between these two is a matter of further debate. Initially it was thought that 70 per cent of the emissions reduction would be met by carbon compliance and 30 per cent by allowable solutions. An excellent paper by the Zero Carbon Hub[5] highlights that this still remains very demanding and unrealistic for some house types. The recommendations in that report suggest the following carbon compliance levels:

- 60 per cent for detached houses
- 56 per cent for attached houses
- 44 per cent for low-rise apartment blocks.

The carbon compliance percentage achieved on any project will be determined by the efficiencies achieved by the building fabric and building services plus any on-site low- or zero-carbon energy source. In the original proposal for the zero carbon standard there were no specific building fabric efficiency targets. This was addressed in the (2010) *Code for Sustainable Homes: Technical Guide*, which set a new criterion, the Fabric Energy Efficiency Standard (FEES), for the dwelling's space heating.[6] Since the zero carbon standard for new homes is not to be enforced until 2016, FEES is likely to be developed further before then. At the time of writing, it helpfully introduces a Passivhaus-style space heating energy target – for the first time moving away from a carbon emissions measurement. We feel this is a positive development. The lower the space heating target you can achieve, the less you need to make up to meet the overall carbon compliance percentage. Reducing the on-site carbon emissions through excellent fabric performance is also key to reducing the need for additional and expensive on-site energy production. Energy generating systems will also have a shorter expected life (a solar hot water system, for example, will last 10 to 25 years) compared with the general building fabric (average 60 years), so prioritising investment in fabric makes good sense.

For the zero carbon level, the FEES target is (currently) 39-46kWh/m².a. The range reflects different house types – for example, it is easier to improve performance on a mid-terrace than on a detached house. The UK's Standard Assessment Procedure (SAP) method for measuring floor area is more generous than that of Passivhaus (see Chapter 7, page 94), which means that the zero carbon target is actually equivalent to over 50kWh/m².a in Passivhaus terms. The consultation[7] for the next updated Building Regulations, Part L (2013) is also proposing an interim Target Fabric Energy Efficiency (TFEE) requirement for new dwellings, of 43-52kWh/m².a, adopting the same measuring standard as the zero carbon FEES levels. As yet there is no final decision on the standard they propose to adopt – the TFEE or full zero carbon proposed FEES levels. Whichever is adopted will then also link into Level 4 of the CSH. The interim TFEE would apply from 2013 until 2016.

Fabric performance and ventilation strategies

The zero carbon space-heating target range is clearly less stringent than that of Passivhaus (at 15kWh/m².a), and this reflects a reluctance to limit ventilation strategies to MVHR. There is a body of opinion that advocates natural ventilation solutions such as **passive stack ventilation** (a whole-house ventilation system that uses naturally occurring pressure differences to draw air in through trickle vents in windows and then up and out through ducts in the kitchen and bathrooms). This necessitates higher air-leakage rates, which come with a significant energy penalty. The air-leakage benchmark for zero carbon is being currently mooted at an air permeability of $3m^3/hr/m^2$. This is approximately equivalent to 3ach (air changes per hour) for an average-sized house (see Chapter 9 for more details). There has also been discussion regarding the capability of the general construction industry to achieve very low air change rates, i.e. below 3ach. This compares with the Passivhaus standard of 0.6ach – again, a far more stringent standard.

Without mandatory fabric energy targets, there was an early tendency for those looking to meet the zero carbon standard to rely on relatively complex and hard-to-maintain low-carbon heating solutions to achieve the carbon emission reductions. This concentration on renewable technologies can easily lead to a less informed focus on building fabric, which will increase other risks. By beginning to increase levels of insulation in our homes and making them much less 'leaky', we are changing the way they physically behave, and an understanding of this is key. Making radical changes without understanding carries four major risks:

- reduced indoor air quality (IAQ)
- moisture within the fabric causing mould and deterioration of materials (and exacerbating the risk of reduced IAQ)
- unacceptable overheating in summer
- underperforming buildings (in energy terms).

Introducing FEES will – rightly – refocus on the building fabric, but this will need to be coupled with appropriate modelling methods and

Mechanical ventilation and air quality

The debate about naturally ventilated and mechanically ventilated solutions for houses primarily pivots on which will deliver better indoor air quality (IAQ). There are, thankfully, a growing number of studies being undertaken to provide solid data on this important issue. In light of the data available to date, we are convinced of the very real benefits of a controlled mechanical ventilation system – as long as it is designed and installed appropriately. For those interested in this subject, we would recommend the 2012 report of the Ventilation and Indoor Air Quality Task Group set up by the Zero Carbon Hub, entitled 'Mechanical ventilation with heat recovery in new homes',[8] which contains an extremely useful review of the studies on links between IAQ and health. In its introduction it notes that "at low permeability levels, reliance cannot be placed on the ability of the home to ventilate itself". This is why MVHR is an essential component of Passivhaus, which sets the lowest air permeability standard (0.6ach), and why the zero carbon standard proposes a much higher air permeability standard of $3m^3/hr/m^2$ and then allows other 'natural' ventilation solutions – which in turn necessitates less stringent space heating targets.

training so that these changes in building construction do not lead to the risks just noted. Linking the required measurements to as-built performance (measuring real performance after occupation), rather than design performance (using energy modelling software during design), as now recommended in the zero carbon consultations, will assist with this.

A zero-carbon Passivhaus

There is a misconception that Passivhaus "will only get you to L4 (Level 4) of the CSH". If you built a Passivhaus and chose not to address any of the other sustainability criteria assessed in the CSH, then this might be true. In particular, Passivhaus does not include the water usage criteria demanded by L5 and L6. However, if water usage is addressed in a Passivhaus design, it should comfortably meet L5.

In fact it should be easier and cheaper to meet a zero carbon standard in a Passivhaus than in a structure that is not designed as a Passivhaus. This is simply because the building fabric of a Passivhaus leads to the exceptionally low energy requirement for space heating, even relative to a zero-carbon house (see Table 1.1, page 18). The Passivhaus standard also includes a total primary energy demand of $120kWh/m^2.a$, which will also help to achieve zero carbon because it encourages efficient energy use across all electrical appliances and uses within a building.

Together, this means that if regulated and unregulated carbon emissions are taken into account, a Passivhaus can reach a zero carbon level with minimal additional renewable devices. If regulated emissions only are considered (the current zero carbon definition), a Passivhaus has been shown to achieve a 65- to 70-per-cent reduction in regulated carbon emissions compared with a compliant Part L (2006) dwelling, when calculated using the PHPP, without the use of any on-site low- or zero-carbon energy provision.[9] This meets all the current zero carbon 'carbon compliance' recommended emissions reductions (44, 56 or 60 per cent). In making a comparison between Passivhaus and zero carbon, it should not be forgotten that, by using the PHPP and applying Passivhaus methodology, a much more reliable prediction of real-life performance is achieved. And, as we discover later in this book, Passivhaus also addresses summer overheating risks (see Chapter 11) and IAQ (see Chapter 12) much more reliably than a non-Passivhaus low-energy building.

Even where a project aim is to achieve zero carbon, it is worth giving serious consideration to using the PHPP and applying Passivhaus methodology and principles to the building fabric design, i.e. to meet or, even better, to exceed, the FEES target. There are many efficiency gains and no conflicts.

Broader sustainability criteria

As we have seen from the issues relating to ventilation strategies discussed above, the Passivhaus focus on building fabric performance, and scientific research into and testing of this, is critically important. The fact that other sustainability criteria are not included as part of the Passivhaus standard has ensured that this focus has been maintained. This is not to say, however, that other sustainability criteria apart from carbon emissions (which zero carbon focuses on) and energy in use (which Passivhaus focuses on) are not important: the fact is that Passivhaus buildings can be made from many different materials and construction methods (both lightweight and heavyweight), and the Passivhaus standard is perfectly suited to combining with more diverse sustainable assessment systems. Unfortunately, official certification using two different assessment methods on one building will have cost implications, but in technical terms there is no inherent incompatibility between Passivhaus and other systems, such as the CSH or BREEAM, or the US systems LEED and HERS. Other standards can usefully widen the Passivhaus approach to consider some broader sustainability issues, including recycling, water management and 'Lifetime Homes' recommendations.

On-site low- or zero-carbon energy

While a zero-carbon Passivhaus may not require any or only minimal on-site or zero-carbon energy for carbon compliance, you may still want to consider such options. If the aim is only to meet the zero carbon FEES targets (not the full Passivhaus target), then some on-site or zero-carbon energy will be essential. ('Zero-carbon energy' normally refers to biomass fuel, while 'on-site' refers to energy generation.)

Offsetting carbon emissions from electricity use

As can be seen from Table 5.2 overleaf, UK grid electricity is carbon-intensive relative to other fuels used in the UK. While it is possible to construct a building that is not connected to the grid, this is not often a practical option and it would certainly make no sense, economically or environmentally, to deploy off-grid solutions at any significant scale, as long as there is a functioning electricity grid! Most homes are and will be grid-connected, so unless and until the grid itself is supplied entirely from non-carbon-emitting sources, generally a zero-carbon building will need to offset 'dirty' grid electricity by photovoltaic (solar-generated electricity) panels. Take, for example, a building located in southern Britain with an unshaded roof of sufficient area, appropriately oriented, and an installed solar photovoltaic array of sufficient size – say, 4kWp (kilowatt peak). Over the course of a year, this would generate electricity equivalent to the annual consumption of a reasonably frugal average-sized household (say, 3,200kWh per year), assuming no electricity is used for hot water or heating.

Table 5.2 **Greenhouse gas content (CO_2e) in kg per kWh of delivered energy for various fuel sources**

Energy source	CO_2kg/kWh
Grid electricity	0.525
Coal	0.333
Gas oil	0.267
Fuel oil	0.267
Diesel	0.252
Burning oil	0.247
Petrol	0.241
LPG (liquefied petroleum gas)	0.215
Natural gas	0.184
Wood pellets	0.039

Figures are as at 2011, on a gross calorific value basis. Source: Carbon Trust / Defra[10]

Space heating and hot water

In summertime, water can be heated using solar thermal panels. Installed on an appropriately oriented roof, they provide zero-carbon heat, assuming the electricity needed for the pumps is also zero carbon.

Wintertime hot water and space heating can be provided from biomass fuel, which can be zero carbon (generally, 'biomass' refers to any organic solid fuel, but in the context of space and water heating here, we use it to refer to wood in pellet, woodchip or log form). In a conventionally built zero-carbon house, biomass fuel would typically be burned in a physically substantial woodchip or wood-pellet boiler; preferably one with a lower output than that required for a standard house. In a Passivhaus, a similar biomass boiler could provide enough heat for a small apartment block or a terrace of Passivhaus homes via a **heat main** (a system of insulated pipes that run between buildings). Woodchips and wood pellets are not truly zero carbon, because the processing and transporting of chips or pellets almost always requires some fossil fuel input, but they are nevertheless very low-carbon fuel sources.

In a conventionally built zero-carbon house, a log burner with a back boiler would deliver zero carbon heat, if the wood was located nearby and had been cut up with an axe! In a zero-carbon Passivhaus, the problem to date has been that there have been few wood log burners or other types of biomass boiler with a low enough space heating output to make them suitable for a Passivhaus. However, this is beginning to change as the first Passivhaus-suitable models come on to the market. For people living in rural areas, it can make sense to use a low-output wood burner as a heat source.

However, large-scale wood burning for home heating is not practicable because, even with much-improved woodland management, the UK's wood supplies are insufficient to meet the large demands that our poorly insulated housing stock would place on them. If all UK homes had been built to Passivhaus standards, it would be much easier to provide the small residual amount of heat they would need with a well-managed wood resource. This theoretical scenario is not very realistic in the UK, with its large, densely populated cities, because it would require the transport of large quantities of solid fuel from the countryside, and the concentration of wood burning would be detrimental to local air quality.

The final point to remember in relation to all these space- and water-heating options is that, unlike the fabric of the building itself, they do not for the most part represent investments that will survive for the lifetime of the building. New builds are typically designed for a 60-year lifespan. Passivhaus buildings should last longer owing to the low air leakage rates and the Passivhaus focus on build quality. Over the lifetime of a Passivhaus, space heating and hot water provision will probably need to be replaced several times (just as in a conventional house). So there is no guarantee that a zero-carbon Passivhaus (or any other zero-carbon house) will remain so. For example, a notionally zero-carbon wood-pellet boiler can be replaced by a boiler run on fossil fuel, such as natural gas. In contrast, it is much harder to change the fabric of the building and 'break' the building's original design intent.

Passivhaus and post-peak energy supplies

While public policy in many European countries, including the UK, has focused on responding to climate change, the UK government has singularly failed to accept, let alone attempt to respond to, the separate and equally intractable predicament we face with energy. As we noted in Chapter 2, energy supplies are very likely to be expensive and ultimately less reliable in the coming decades, and this will impact on the economics of Passivhaus as well as on other forms of building. Two concepts that are key to this issue are '**net energy**' (or 'net energy gain/balance') – the remaining energy available to society after the energy needed to obtain it from an energy source has been subtracted – and '**peak net energy**' – the maximum rate at which net energy can be extracted from a source. Peak net energy is a more accurate term than the more commonly used 'peak oil'. It refers to the rate at which we are able to extract net energy from our environment, and it is fundamental to our success as a species. Net energy can be expressed as a ratio: **energy returned on energy invested (EROEI)**[11] (sometimes referred to as 'energy returned on investment'; EROI) – the number of units of energy produced for each unit of energy consumed in order to produce it. Figure 5.1 below shows a simplified representation of EROEI.

The fossil-fuel era started with the large-scale mining of coal in England in the eighteenth century and grew even faster as oil production began in earnest, commencing in the United States. Globally it heralded a period of high EROEI or high net energy, and this 'easy' energy has given the rich world a material standard of living unprecedented in all human history. The flood of energy has also allowed us to create a society of extraordinary complexity and specialisation, again unprecedented in all human history. We have had access to so much energy that society has had incredible freedom to decide how that energy is used. Oil is the most important energy source of all, as it is an enabling or master energy source that allows us to unlock others – take oil out of the system and virtually all other forms of energy production are severely compromised. Oil is critical to our transport systems; indeed transport is the hardest sector to decarbonise, because our

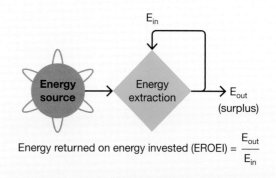

Figure 5.1 Energy returned on energy invested. *Source: Hall et al. (2002)*[12]

transport infrastructure is designed to operate with liquid fuels.

While there has been periodic concern about oil 'running out' – notably in the aftermath of the 1973 oil crisis – it has become clear that natural resources do not generally suddenly run out; instead, the rate of extraction reaches a maximum or peak before going into decline. Technical advances can extend the tail of the decline, and other human interventions, such as geopolitical factors, can reduce the total amount ultimately extracted. The EROEI of the energy extracted before the peak is a lot greater than that of the energy extracted post-peak. Globally, we are approaching, are at or have passed a peak in net energy from oil extraction. There has been much debate about when this will occur, or whether it has already occurred, but few question the fact of a peak in production. Unfortunately, changing infrastructure is itself very energy-intensive, so there is no 'silver bullet' that will suddenly transform our energy future. As EROEI drops below 8:1, society becomes far more constrained by its energy choices.

Joseph Tainter, in his seminal work *The Collapse of Complex Societies*,[13] and others have examined how societies have historically adapted to a shrinking EROEI, i.e. a decline in net energy. In simple terms, societies past and present have used their ingenuity to increase the efficiency with which they employ their resources to counter the impact of reduced net energy flow rates. This is very much the path our global society has taken over the past few decades. For example, growth in computer and Web-based systems has allowed increased economic activity for a given energy input. The downside of this trend is that our productive systems have become much less able to cope with quite short interruptions in the energy supply. Supermarkets, banks and supply chains almost everywhere rely on 'brittle', just-in-time inventories:

Transitioning to a lower-EROEI future

EROEI figures are sensitive to many assumptions and variables. Historically, the production of oil, coal and natural gas had EROEIs of between 100:1 and 20:1; the EROEI of renewables ranges from around 35:1 to little more than 1:1! (Biofuels produced in temperate climate zones, seen as a panacea in some circles a few years ago, have an EROEI close to 1:1.) But it is evident that energy generation based on renewables provides far less net energy than fossil fuels have done in the past. And the EROEI for fossil fuels is also declining, as extraction becomes more difficult and the 'low-hanging fruit' is mostly gone. It is clear, then, that we need to reduce our overall energy demand in order to adapt to the lower-EROEI energy sources that will be available to us in the future. To help achieve this, high-performance buildings in terms of energy invested for energy conserved make the most sense.[14]

if any part of the supply chain is unable to deliver, even for a few hours or days, the impact is disproportionate. In these cases, efficiency has been achieved at the expense of system resilience.[15]

How resilient is Passivhaus?

A Passivhaus build, like any conventional building project, is dependent on reliable energy flows – energy is not just needed on-site; obviously, it is also needed to manufacture and shift building materials and to transport the build team. The specialist components used in the first UK Passivhaus builds typically travelled around 1,000km from factories in central Europe to the UK. These distances should shrink over the coming years. Already more companies are producing suitable products, and even UK manufacturers have started responding to the demand for much higher-performance windows and Passivhaus-certified MVHR units. However, fast-forward a couple of decades and it may prove impossible to get timely delivery of bespoke specialist components such as windows. The rationing of liquid fuel – either by price or via a rationing scheme such as tradable energy quotas (TEQs)[16] – will make local and regional products comparatively more attractive and feasible. Whether MVHR units will be produced on a local or regional basis is hard to say. However, these present less of a problem as they could be pre-ordered by regional stockists.

As we suggest in Chapter 14, part of the policy change needed to facilitate growth of Passivhaus in the UK is to encourage domestic manufacture of Passivhaus building components and to make changes to the structure of the construction sector, including training and skills (see also Chapter 4). These changes would also help to make Passivhaus a more resilient approach, particularly if using more low-impact building materials. In other words, there is nothing about Passivhaus that makes it intrinsically less resilient than other forms of building in a lower-net-energy future. Nevertheless, clearly if the net energy available to society drops below the levels required to maintain a functioning industrial society, we will be living in a world where more immediate needs – food and security – take precedence. It is not possible to predict exactly how our energy predicament will play out. It depends, in large part, on the wisdom and far-sightedness of the policy decisions we have taken, are taking now and will take in the future.

RECAP

There is no innate conflict between Passivhaus principles and a preference for building with natural and low-embodied-energy materials. Some elements of a Passivhaus will necessarily need to be less 'natural', in particular where specific performance criteria must be met, e.g. for windows or ventilation systems, but this will be true for most buildings. There is also no innate conflict between 'zero carbon' and Passivhaus; in fact, we would recommend that the Passivhaus methodology should be adopted to efficiently and accurately meet or exceed the zero carbon FEES targets. Passivhaus can help to meet carbon compliance with little or no reliance on on-site low- or zero-carbon energy. Furthermore, by opting for a mechanical ventilation strategy, good levels of indoor air quality will be more readily achieved. In turn, other assessment standards can usefully widen the Passivhaus approach to consider some broader sustainability issues.

In a more energy-constrained future, Passivhaus has the potential to offer a resilient housing solution, not least because of its efficiency and because it does not rely so heavily on short-life complex technology, although there is a need for more localised supply chains for some of the Passivhaus components.

CHAPTER SIX
Setting up a Passivhaus project

Choosing a plot, planning considerations, retrofit considerations, phased retrofitting and extensions, selecting an architect and builder, the role of the client

This chapter is mainly of interest to those wishing to commission a new-build or retrofit project. While some points made here relate specifically to new builds, retrofits or extensions, many are relevant to all Passivhaus projects. There are many variables to consider when setting up a building project, but we have restricted the discussion here to those that can have an impact on Passivhaus builds.

Choosing a plot

Acquiring a viable building plot at a sensible price in a half-decent location is a challenge in the UK. Any choice will involve uncomfortable compromises. That said, it is worth examining the factors that will impact on a future build, particularly a Passivhaus. Similarly, it can be a challenge to find a building that lends itself reasonably easily to being retrofitted.

There are many considerations involved in choosing a plot that apply to building projects in general and are outside the scope of this book; these are covered by other publications and online resources. However, one worth mentioning here is site access. Plots with difficult site access or steep inclines will cost more to build on, and this is just one of many factors affecting cost (other factors were discussed in Chapter 2). If you are able to choose a plot that is cheaper to build on, you will of course be freeing funds for the build. If there are concerns about site access, it's a good idea to ask a builder for an opinion.

None of the issues discussed in this chapter will determine whether or not the Passivhaus standard is achieved, but they will have a bearing on the cost of the project. If you already have a plot, you may be constrained by the extent to which you can change these variables, depending on how much scope there is in determining the building's location on the plot.

Shading

As we will see in the next chapter, during wintertime, solar gain – heat energy gained passively through the windows – makes a significant contribution to reducing the building's heat demand. If little or no wintertime solar gain is available, it means that the roof, walls and floor U-values will need to be improved considerably to bring the overall net heat demand down. Figure 6.1 overleaf shows the energy losses and gains through different elements and from different sources in a typical Passivhaus. The amount of energy needed to balance the building's heat losses and heat gains – the heat demand – is 15kWh/m².a in this example. **Energy balance** is a key Passivhaus concept.

Check whether the location of your proposed new build or retrofit building is in shadow due to buildings, trees or other structures adjacent to your plot. This is particularly important if the shading structure or tree belongs to a neighbour or if you would like to retain it yourself. Take care not to overlook any trees with tree preservation orders (TPOs) or other planning restrictions that might have an impact on shading, now or in the future. If in doubt, check with your local planning authority. Your solicitor should also highlight any TPOs on the plot, if you are in the process of buying it.

Some sites may have east or west shading, which can be beneficial in helping to reduce summertime overheating risk.

While some shading will no doubt be unavoidable, a judgement will need to be made about acceptable levels of shading. Chapter 7 explains

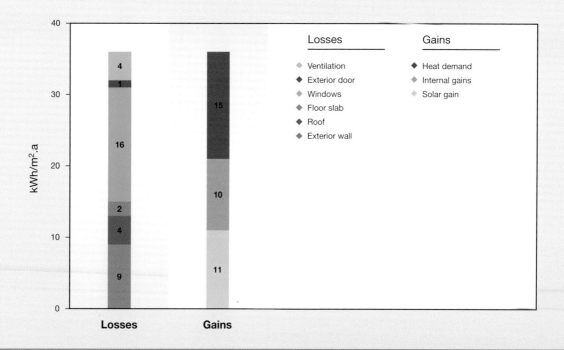

Figure 6.1 Energy losses and gains, and overall energy balance, in a typical Passivhaus.

how shading is assessed in the Passivhaus Planning Package (PHPP). If you have already chosen an architect with experience of Passivhaus (see page 83), he or she should be able to provide some advice on shading, as well as on other issues related to choosing a plot.

Planning considerations

If you are considering a retrofit project and your proposed purchase is a listed building, or is in a conservation area or national park, making changes to the external appearance of the building will be much harder, if not impossible, and therefore likely to exclude the usual option of external wall insulation. If a new-build plot is located in a conservation area or national park, it will also be subject to more restrictions. However, even if this is the case, a new build does not have the constraints of working around an existing structure and so allows much more discretion in the internal make-up of the building assemblies, as long as the external appearance is deemed to be 'in keeping' with its neighbours.

Conservation areas and Passivhaus components

Although it is technically possible to reach the Passivhaus standard by insulating internally, doing so will involve more costly and invasive interventions inside the building, which can often compromise the internal layout and will nearly always reduce useful floor area. Internal insulation also requires much more careful attention to moisture management, to avoid damage from cumulative build-up of water and water vapour within the building structure. For most buildings, being located in a conservation area – especially if the building has listed status – will impose additional costs and restrictions that will, in many cases, make reducing its energy use down to Passivhaus or even EnerPHit standards economically unfeasible, and probably too risky from a moisture-management perspective. However, Passivhaus methodology can still be applied pragmatically to get the biggest improvement in energy performance within the practical constraints of the building and the financial constraints of the project. Finding a suitable 'conservation-friendly' window product that will perform adequately for an ultra-low-energy building is difficult; until recently there were few, if any, such products on the market. In the Princedale Road retrofit project (pictured top right, highlighted in pink), a bespoke window solution had to be designed and developed by the contractor, EcoHaus Ltd, in order to meet the requirements of both the local conservation officer and the Passivhaus standard, as no suitable products had been brought on to the market at that time. A wider range of styles and Passivhaus-certified windows is now available, an example of which is pictured below right.

Even if your proposed purchase is not listed and is outside a conservation area or national park, if it is part of an estate or a row of relatively homogeneous properties that has a group value (in terms of architectural merit), it will probably be subject to more constraints than a building that is not. There may also be other circumstances that affect the degree of planning scrutiny for a given location.

It is worth noting that many planning applications make claims about their sustainability credentials to help the application gain planning permission; this is particularly true if 'sustainability' is being used to argue a case to build in a location where planning permission would not otherwise be granted. Planners are aware of this

Princedale Road, West London: the UK's first retrofit project to achieve full Passivhaus Certification. *Image: Paul Davis + Partners*

Wooden Passivhaus-certified windows by ENERGATE® in a period building in Germany. *Image: ENERGATE®*

tactic, so be sure to explain (briefly) what Passivhaus means and its wider significance. If your project is going to be certified, mention this and its significance.

Seeking planning advice

There are many other factors that the planners will take into account, which are not Passivhaus-specific but affect all developments. Further reading is recommended (see Resources), to allow you to assess potential plots through the planners' eyes. If the project size and budget warrant it, or if planning is likely to be particularly challenging because of factors specific to your chosen location, it may be helpful to hire the services of a planning consultant. Planning is discussed in more detail in Chapter 14.

Retrofit considerations

In the case of a retrofit project, there are further issues to consider, as follows. Even if the building or plot adequately addresses all of the planning requirements, some buildings lend themselves better to retrofitting than others.

- **Original brick or stone façades** These can be externally insulated, but there are two issues. First, board-based insulation will require that the original façade be rendered before application of the insulation, adding to the cost. Second, while it is possible to buy external insulation products with a false brick façade, they really do tend to look false(!), and your building is likely to look better if you chose not to attempt to recreate what was there before.
- **Cavity walls** These create two problems. They are often thicker than their solid (single-skin) wall equivalents, so the addition of sufficient insulation to achieve the Passivhaus performance results in overly thick walls (Passivhaus discourages wall depths beyond 500mm). On many sites, such thick walls are not practical because of space constraints, nor desirable for aesthetic reasons. Also, the cavity itself is a problem if left unfilled, because air movement within the cavity causes heat transfer by convection (known as **thermal bypass**). Of course, many cavities are already filled, and in some of these the existing insulation performs poorly and may be poorly installed (see Chapter 8, page 112). However, this doesn't mean cavity walls present an insurmountable problem – the Totnes Passivhaus, for example (pictured on page 53), is a cavity-wall retrofit.
- **Concrete slab floor combined with low ground-floor ceiling heights** If your building sits on a concrete slab, it will probably not make economic sense to remove it, so floor insulation will need to go on top of the slab. If there is already limited ceiling height and the finished floor level therefore cannot be raised, then unless the first floor is to be removed (a huge intervention by any standards!), the space for floor insulation will be very restricted. While this problem might be addressed by using specialist ultra-low-thermal-conductivity insulation, such as vacuum-insulated panels (VIPs), aerogels or similar, this is particularly expensive.
- **Suspended wooden floors** Many homes have suspended wooden floors in all parts of the ground floor except the kitchen, where there is often a concrete slab. In this case it can be fairly straightforward to remove the slab, as it is probably not a structural element supporting the building. Insulating suspended floors is relatively straightforward if there is enough space to access them from below. This is particularly helpful if you want to retain the existing finished floor – a parquet, for example. However, the majority of suspended floors have only a 350-500mm

void, making insulating from below either very difficult or impossible, in which case insulation would need to be installed from above. But removing the existing floorboards also has its benefits, as it makes it much easier to properly address airtightness requirements and deal with joist ends. The insulation of suspended wooden floors has to be detailed carefully to ensure that moisture will not form and collect around the joist ends. Some would favour removing the suspended floor completely and replacing with a solid concrete slab, as it's quicker (although messier) and makes it easier to detail for insulation and airtightness continuity. (For more on managing moisture risks when insulating, see Chapter 10.)

- **Roof overhangs** A building with deep roof overhangs should make it simpler to insulate externally because the roof does not need to be extended outwards to accommodate the insulation. It is important to be able to remove existing soffits (the underside of the roof overhang), as this allows the new external insulation to extend upwards to meet roof insulation (warm roof construction) or loft insulation (cold roof construction), minimising the thermal bridge along the wall–roof junction. For the latter, this building junction needs to be detailed very carefully to ensure there are sufficient airflows in the loft space.
- **Roof ridge line** If tied into a fixed ridge height (i.e. it's not possible to raise the ridge owing to adjacent buildings, for example in a terrace), your roof insulation solution will be less simple and this might drive you to insulate across the top-floor ceiling instead ('cold roof construction'). Normally, Passivhaus insulation follows the roof line.
- **Fireplaces** These are not compatible with the Passivhaus approach, as they are designed to draw volumes of air (creating draughts) through your building in a very energy-inefficient manner. Blocked-up flues can become potential moisture traps and need to be ventilated, which again is counter to low-energy design. It is perfectly possible to retain external stacks supported on gallows brackets, but it is best to remove chimney breasts internally. There are ways to seal a

A building with a generous roof overhang, such as the one pictured right, makes it easier to insulate externally than when there is virtually no roof overhang, as in the photo on the left.

chimney flue and provide a Passivhaus-suitable log burner, but this would need to have a specifically controlled air supply and exhaust, sealed to the room, which has cost implications.
- **Windows** Some window designs will be more problematic and expensive to upgrade, e.g. dormers and bays.
- **Extra complexity** The time and money needed for design and energy modelling is greater in a Passivhaus retrofit than in a new build. A structure that was not designed to be to a Passivhaus creates specific issues that need to be overcome, often by quite creative but time-consuming solutions.
- **Form factor** The more spread-out your building, the harder it will be to achieve the Passivhaus or EnerPHit standard (see Chapter 1, page 21), and in some cases it will be impossible.
- **VAT** As with all refurbishments in the UK, the full VAT rate is payable for retrofit build costs, whereas it can be claimed back in a new build. This anomaly does not apply in Ireland. With VAT currently at 20 per cent, the ex-VAT cost of the retrofit will have to be a lot less than that of an equivalent new build for it to add up financially.
- **Thermal bridges** Nearly all retrofits are going to have a thermal bridge at the floor–wall junction that cannot be eliminated practicably or economically. Further thermal bridges are likely at the wall–roof junction. Any thermal bridges in a building approaching the energy efficiency of a Passivhaus will add significantly to the overall heat load. The heat loss arising from each thermal bridge has to be quantified in separate thermal bridge heat loss calculations; it costs upwards of £250 per junction to calculate its specific heat loss or **psi-value** (the measurement of heat loss in a linear thermal bridge). In a new build, it is possible to design out all thermal bridges, making this additional work unnecessary. (See Chapter 8.)
- **Ventilation systems** Installing a visually discreet ventilation system will often be harder in a retrofit because of constraints imposed by the existing structure. There will probably not be enough room for ducts between internal walls or within internal floors (where ducts need to run perpendicular to the floor joists). Ducts in a Passivhaus are oversized to reduce air speeds; slower air speeds improve energy efficiency and reduce noise. (Chapter 12 covers ventilation system design in more detail.)
- **The quality of the original building** Although it is sometimes hard to judge accurately until after retrofit work has commenced, it is vital to take time and get professional advice on whether a building is worth saving before deciding to go down the retrofit route. It is easy to let a (quite reasonable) belief in the value of retrofitting unduly influence your decision. There are many buildings in the UK that were built very poorly and really need to be put out of their misery! If building materials, particularly those with high embodied energy, can be viably reclaimed, all the better. Building new makes it much easier to reach the full Passivhaus standard and much more practical to use low-embodied-energy materials. Space constraints in a retrofit very often force the designer to choose between higher-performance but usually high-embodied-energy materials, such as foam board insulation, or lower-energy-performance but more environmentally benign materials. In a new build it is possible to do both.

Phasing retrofit work

Homeowners and registered social landlords (RSLs) may not be in a position practically or financially to undertake a full Passivhaus (EnerPHit) retrofit in a single phase. Such work almost always requires that the building is vacated for the duration of the build, and committing to such a large capital spend can be too big a burden, especially in the era of tighter credit conditions. The alternative is to plan a retrofit in stages.

The sequence of phased retrofit work may well be driven by the cycles of planned (or unplanned!) building fabric repair or replacement work. The common-sense approach is to plan strategically, taking into account how earlier work phases will impact on the cost of planned future work phases. This reduces the risk of having to undo earlier work. If the strategic goal is to get the building to perform to the EnerPHit standard, each phase will need to be approached as if it were part of a larger EnerPHit retrofit. For example, if working on the repair or replacement of a ground floor, you would not only install the required insulation; this work is also an opportunity to install an airtightness layer in the floor and to plan and execute the installation of insulation to minimise thermal bridging. These steps are vital in managing moisture risk properly as well as improving the overall energy performance (U-value) of the floor.

A similar approach could be taken when insulating or repairing a roof. Careful thought would be needed when planning the path of the airtightness layer, particularly where the roof meets the external walls.

Upgrading windows is more of a challenge. It is essential that the strategy for installing windows takes into account how the walls will be insulated. The Passivhaus retrofitting approach would by default involve external wall insulation. New windows would then be installed, mounted on the outside of the existing walls, with external wall insulation wrapping around the window frames to minimise thermal bridging. If the windows are to be installed months or even years before the external wall insulation, the windows will have to be installed in such a way that they work practically (no gaps around the edge, properly weather-proofed, etc.) in the interim.

To do cost-effective phased retrofit work, where the aim is to eventually achieve the EnerPHit standard, before any work starts on-site you will need professional design input by a Passivhaus Designer, PHPP energy modelling and possibly moisture modelling (see Chapter 10, page 158). In design terms, the project should be treated as if it were being undertaken in a single phase, with the caveat that the design needs to consider potential moisture risks between each phase of the work. An architectural detail of a building assembly or junction could cause moisture problems between interim construction phases, even though it may well perform correctly in energy and moisture management terms after the last phase is completed.

Extensions

It is also possible to adopt the Passivhaus building fabric standards for new extensions on existing properties. There is not a great deal of sense in losing the opportunity of building to this level (or near this level) if you are already expending financially on new work. Extra-over costs should be minimal.

This approach would bring particular benefits where the new extension is to be open to the existing house. Here you will generally be removing old, substandard external walls and, in effect, replacing them with new, high-performance ones. If old doors and windows are being removed, then the new extension replacements also offer the potential for an excellent upgrade. It is very common to create new, open-plan living/kitchen rooms when making such building alterations, and these modern spaces will often be where the majority of time is spent.

Janet Cotterell and Christine Harrison's extension, built to Passivhaus standard, in north-west London.

A more comfortable indoor environment can then be achieved while reducing energy bills, plus there will be knock-on improvements to the rest of the house. Such an extension does not require any special ventilation strategy, as the existing house will be more than 'leaky' enough. In the case of a single-storey extension, you could then keep the first-floor sleeping zone at a cooler temperature – which many people prefer. If you are building a large two-storey extension with a great deal of new accommodation, the interlink between old and new will be more complex and the ventilation strategy will need more careful consideration.[1]

You may consider using lower-performance windows (with a U_w of around 1.2W/m²K – see Chapter 11, page 171) if your budget is really stretched; however, there will be some risk of a comfort penalty. Also, if the extension is part of an EnerPHit retrofit that is being built in phases, you must specify Passivhaus-suitable or -certified windows to ensure that the building reaches the full EnerPHit standard after completion of the final phase.

Selecting an architect and builder

The decisions as to which architect and builder to appoint are those with the most far-reaching impact on any build project. In a Passivhaus project, these decisions are even more critical to its success. As discussed in Chapter 4, Passivhaus projects demand a much greater attention to detail during both design and construction than comparable non-Passivhaus builds.

Clearly, it will be a less stressful project for all if the architect and builder see Passivhaus as a goal they want to achieve for themselves; the team should have 'common purpose' (see box opposite). Also, there must be a genuinely cooperative and trusting relationship between architect, builder and client. The importance of cooperation and trust as success criteria cannot be underestimated on a Passivhaus project. It really is worth doing everything possible to make the right choice of people for these key roles, so do allow enough time to reach a well-considered decision.

> ### What is common purpose?
> "Common Purpose is a shared intention to achieve a shared goal, where collective aims are advanced by the individual purpose, and individual aims are advanced by the collective purpose."
> **David Fleming**[2]

Defining the brief

Having decided to go ahead with a build project, your starting point as the client is to write down your aims and requirements in an architect's brief. The Royal Institute of British Architects (RIBA) provides good advice on how to brief architects in its booklet *A Client's Guide to Engaging an Architect* (see Resources). It suggests providing information in the following categories:

- functions of the building
- (client's) motivations and expectations
- design direction
- authority for decision-making
- timetable and budget.

Kevin McCloud's *Grand Designs Handbook*[3] also has some very good tips about how to select and work with your architect.

All the above points apply to self-build projects in general. A brief for a Passivhaus build obviously needs to state clearly that achieving the Passivhaus standard is a key goal for the project, and the criteria for meeting the standard (see Chapter 1) should be stated in the brief. This is important, as it will help to identify which architects really understand what Passivhaus entails. If you are planning to get the formal Passivhaus Certification, you also need to state this in the brief.

Selecting an architect

Once you have your brief, the next job is to find an architect who is able and willing to design a Passivhaus. While RIBA's website (www.architecture.com) will help you get a list of candidates, it is worth looking at two other sources of information to narrow down your list:

- The Passivhaus Trust (www.passivhaustrust.org.uk) – the UK's leading organisation working to promote and protect the Passivhaus standard. Its list of members includes architectural practices and design and build companies offering Passivhaus design services.
- PassivhausPlaner.eu (www.passivhausplaner.eu/englisch/index_e.html) – a European register of Certified Passivhaus Designers and Consultants. Many Passivhaus architects, or someone from the architect's practice, will have taken the Certified European Passive House (CEPH) course and should be listed on this register. This will give you more confidence that the practice understands the methodology and the challenges of a Passivhaus build.

Another obvious point is that there is a lot to be gained from examining the previous work of the architect and, if possible, arranging a visit; even better if it is possible to meet the occupants. An architect who has already built a Certified Passivhaus should be given serious consideration. This does not mean that architects without previous Certified Passivhaus experience are to be overlooked – there are many projects that have aimed for energy and airtightness standards that were ambitious by the standards at the time. The key thing to establish in this situation is what lessons the architect has learned from the experience.

Once a shortlist of candidates has been drawn up, the next step is to meet them. First, send them the brief. It should give them all the information they need to prepare for an interview. Before any meeting, write a list of fairly open questions that will allow you to test the following.

- Can the architect explain what a Passivhaus is?
- What problems and lessons has he or she learned from previous relevant projects?
- Does the architect integrate low-energy goals into his or her design from day one of the process? In particular, does he or she focus first on optimising the building fabric, rather than on renewables and heating technologies?
- How well does the architect understand the importance of the core Passivhaus concepts covered in this book (form factor, shading and solar gain, thermal comfort criteria, designing out or managing thermal bridging, achieving Passivhaus levels of airtightness, ventilation strategy)?
- What are the principal pitfalls and challenges of a Passivhaus project?
- What specific additional documents are needed from the architect in a Passivhaus project?
- Does the architect know what the PHPP is for and understand its significance in Passivhaus design? Specifically, will the architect ensure that the PHPP is used by a competent person to model the proposed design early on and will it be used iteratively to inform the design process?
- Does the architect understand the role of heat recovery ventilation in Passivhaus? How would he or she address mechanical and electrical (M&E) building services (i.e. heating, hot water, ventilation, electrics and plumbing)?
- How committed is the architect to the goal of achieving a Certified Passivhaus?
- How open is the architect to suggestions and input from the builder?
- How does the architect see the role and selection of the builder in the project?

Finally, take a friend to the interviews so that you can compare impressions, and ask to speak to previous clients.

Selecting a builder

Unlike the architect, who does not have to be local to the project site, if at all possible it makes sense for the builder to be based in the same locality as the build. To do otherwise would not make financial sense in most cases. That said, in the Camden Passivhaus – London's first Passivhaus – some of the members of the build team were 'imported' from Austria. However, the economics of a high-end home in one of London's smarter districts does not reflect the economics of more typical builds. Also, pre-2008, demand for builders often outstripped supply in London. Being the first London Passivhaus, it was a big challenge to find a contractor team that could deliver the airtightness and thermal-bridge-free building standards required in a Passivhaus.

It will be a few years before it becomes realistic to find candidate builders with a track record in building ultra-low-energy and Passivhaus buildings. However, builders who have experience of managing conservation and renovation projects on listed buildings, and who can show that they have a genuine interest in and understanding of low-energy building, are quite likely to be good candidates for a Passivhaus build – as are builders who have previously worked on a project where they have had success in managing airtightness issues without excessive spending on remedial work. A background in conservation helps because such projects share some of the same challenges as builds aiming for good airtightness, because in a conservation project, building work can easily cause unintentional damage to the existing structure. Similarly, airtightness, once achieved in a partially complete building, can easily be compromised during subsequent phases of the build. Builders will need to demonstrate that they have a strategy to manage this. Chapter 9 explores this in more detail.

It is, of course, important to visit sites of builders' previous work and, if possible, to talk to their clients. Try to visit both a complete build and a build in progress. The former should show the craftsmanship of the building team. The latter should, with a careful eye, reveal something of the building team's ability to achieve the very low levels of air leakage needed for a Passivhaus, as well as the team's attention to detail in installing insulation without any gaps, so that no unintended thermal bridges are created.

Visiting a site of a partially completed build – say, at electrical first fix (where the structure is still exposed) – allows you to check whether or not the build quality is just skin-deep. However, the relevance of a site visit will be limited if the builder is planning to use different personnel for your project. Find out whether the foreman or site manager on these previous projects will be available for your project. That person will have a key role in ensuring avoidance of both air leakiness and thermal bridging on-site.

Traditionally, construction contracts are formally tendered to three or four firms using Joint Contracts Tribunal (JCT) contracts (a standard format of contract and associated documents

After completion

The role of the professional team should not end abruptly with the completion of the build. Although a Passivhaus does not rely on a lot of technology and should be straightforward and intuitive to live in, there is a benefit in the architect or developer providing the occupants with a written briefing on the principles underpinning the building and practical tips for achieving the most comfortable, economic use of the building.

The two key areas that will feel unfamiliar to most people are ventilation and shading. An explanation of how to operate the MVHR unit and how to change its paper filters (see Chapter 12) is needed, as well as how to minimise the risk of overheating in summer. Where the design utilises fixed shading devices, this is not as important, but where summertime shading relies on occupants adjusting blinds, some advice is essential.

The Passivhaus standard is increasingly used in social and cooperative housing in Germany and Austria. In these developments, MVHR would normally be managed by the social landlord rather than the tenant, and in fact this is less of a management overhead than dealing with the heating and ventilation infrastructure of a standard building.

Because of this and the very low energy costs of heating a Passivhaus, some landlords are returning to the practice of incorporating heating costs in the rent: a practice common in the former East Germany, where such rents were known as 'warm rent'.

The Passivhaus Institut has produced a series of very detailed 'user handbooks' for occupants and facilities managers.[4] These would need to be translated and adapted for other countries.

used in the construction sector). It may be that you prefer a particular contractor and there are alternative approaches, including 'cost plus'. This essentially means that you pay the actual build costs plus a pre-agreed mark-up. If there is trust between client and builder, this can be an economical approach for the one-off self-builder. It is outside the scope of this book to provide advice on contracts, but an architect should be able to advise.

The role of the client

As has been discussed elsewhere in this book, a Passivhaus is not a bolt-on extra that can be simply tacked on to a non-Passivhaus design. For the client, this means being open-minded about how the build is approached and how the budget is apportioned. A Passivhaus will require money to be spent differently: more on windows and insulation products, without a doubt; plus a bit more for airtightness tape and membranes, and airtightness testing. Funds are also needed to pay for PHPP energy modelling and, optionally, Passivhaus Certification. However, energy modelling should save more in building materials than it costs in fees for the modelling. If the team is new to Passivhaus, certification could also save money by reducing the risk of errors, and it may also help to hold or boost the property's value when selling. Money can be saved if you are flexible enough about the design to take on board advice from the professional team – architect, builder and Passivhaus Designer or Consultant – who should be able suggest ways to make the build cheaper. The form factor is probably the most important variable. This does not mean that inefficient form factors are ruled out; they just have a price tag attached to them (this is of course true for all buildings, not just Passivhaus). There are huge cost variables within a build: finishes (kitchens and bathrooms in particular) can vary by large margins and this does not always reflect quality.

You can help reduce the risk of cost overruns by allowing plenty of time for the design process. The more thought that is given to the project before it starts on-site, the fewer pressurised, last-minute (nearly always costlier) decisions have to be taken (see below). This means that you must be willing to spend a realistic percentage of your budget on design and energy modelling before starting the build. Where the budget is tight, it is tempting to spend the minimum needed on an architect to get a design through planning, then hand the plans over to the builder who has given the most optimistic view of the likely cost. But cutting corners in the design stage is unwise in any project; particularly so in a Passivhaus one. It has two main disadvantages.

First, while it is nice to hear a rose-tinted view of the build costs at the start of the project, this only puts off the day of reckoning, when the tenders or bills come in. For client and builder, not being realistic about the costs from day one will undermine the trust crucial to a Passivhaus project.

Second, insufficient time and resources spent on design before going on-site will inevitably result in 'designing on the hoof' during the build. Design decisions made during the build will be rushed and will increase costs and risks. It is not always clear what multiple effects will result from a design decision, and hasty decisions will therefore lead to unforeseen effects. Most design changes require building materials to be bought or necessitate returns of previously bought materials (if the supplier accepts returns). Last-minute purchases usually do not allow enough time to shop around for the best deals;

sometimes the exact product may not be available and the substitute may not be appropriate or may be more expensive. Furthermore, unplanned changes in a Passivhaus build will add to the risk that corners are cut to make up for lost time. In a standard build, cut corners are often hidden beneath the finishes of the completed building without any apparent consequences, but in a Passivhaus build, those cut corners risk jeopardising the goal of creating a sufficiently airtight and thermal-bridge-free structure. Time on-site is also extended, which has a cost implication. This is not to say that changes on-site can or should always be avoided – sometimes they are necessary – but the temptation to change the design for marginal benefit should be resisted if at all possible. The full costs (financially and in terms of project delays) of making late changes need to be understood at the point the decision is made, or costs will build up over time and there will be a loss of financial control. Changes during the build will likely demotivate the site team, if these involve undoing work already started.

RECAP

Passivhaus projects have many challenges that are common with non-Passivhaus build projects, but there are some additional factors that should be taken into account at the outset. When choosing a plot, shading (both during winter and summer) is a significant variable. Perhaps one of the most important considerations regarding the build is form factor. Planning, particularly if the plot or building is in a conservation area, needs careful attention. While in most cases there are only a few scenarios that are 'showstoppers', it is clear that features of some plots will have an impact on costs.

In a retrofit project, the existing structure imposes many constraints, each of which, if not considered, will add to the challenge (and the cost) of the project. Some people will not be in a position, for financial or practical reasons, to retrofit a house in one stage, in which case it is possible to plan to do the work in phases. Even if it isn't possible or desirable to retrofit, and only an extension is envisaged, there is no reason not to use Passivhaus techniques to optimise the energy performance of the new addition – doing so will bring benefits to the whole building.

The most important success criterion of any Passivhaus project is that of cooperation, trust and 'common purpose' between client, architect and builder, since a first Passivhaus project is likely to present challenges for all three parties. Clearly, knowledge of Passivhaus and its particular challenges is vital, including, for the designer, the importance of investing in early PHPP modelling. Whatever type of Passivhaus project is being planned, careful choice of the design and construction team is especially crucial.

PART TWO

Passivhaus projects:
a practical guide

CHAPTER SEVEN	
Using the Passivhaus Planning Package (PHPP)	**90**

CHAPTER EIGHT	
Thermal bridges	**110**

CHAPTER NINE	
Airtightness and sequencing	**122**

CHAPTER TEN	
Moisture	**144**

CHAPTER ELEVEN	
Windows	**168**

CHAPTER TWELVE	
Ventilation	**188**

CHAPTER SEVEN
Using the Passivhaus Planning Package (PHPP)

History of the PHPP, PHPP worksheets: Verification, U-Values, Ground, WinType, Windows, Shading, Ventilation, Annual Heating Demand, Summer, Shading-S, DHW + Distribution, SolarDHW, Climate

The Passivhaus Planning Package (PHPP) is a design tool to help architects and Passivhaus Designers achieve the Passivhaus or near-Passivhaus ultra-low-energy standard. It takes the guesswork out of the design process by accurately predicting a proposed design's energy performance.

Once a design has been set up in the PHPP, the Passivhaus Designer and architect can test out changes to see their effect on the building's energy performance. This process, used iteratively, allows designers and clients to weigh up different options for trimming back the spec to the minimum needed for the desired energy standard, thereby avoid over-engineering. This can save thousands of pounds, particularly on high-cost items such as windows, even on the build of a single home.

The history and accuracy of the PHPP

The PHPP was first developed in 1998, drawing on experience modelling the first prototype Passivhaus in Darmstadt-Kranichstein in Germany, where the bespoke software models required the entry of thousands of pieces of information in order to get a working model of the design. The PHPP requires a lot less data to be entered but still models the design's energy performance accurately. That said, compared with the Standard Assessment Procedure (SAP) energy assessment tool commonly used in the UK, the PHPP today still demands considerably more time to enter all the data needed to create a working model. However, this extra effort is worth it, because the PHPP has an excellent track record of accurate prediction – as illustrated in Figure 7.1 overleaf, which compares energy use predicted by the PHPP with the 'real-world' results for a range of occupied homes in Germany: a typical 'low-energy' development and three Passivhaus developments. The PHPP models energy use with an assumed internal temperature of 20°C. In an estate or 'settlement' of similar houses, different occupants will have their own ideas of what constitutes a comfortable temperature.

PHPP – science versus art?

Some have expressed the view that the PHPP can influence the design process to its detriment, by cramping the artistic or creative inspiration needed to realise genuinely uplifting architecture. As has been discussed, most build projects are subject to multiple constraints, many of which can or do hinder the creative process. In recent years, energy performance has been added to this list of constraints. By enabling experimentation with different design options, the PHPP lets the designer see the effect of those options on energy performance. In the hands of a creative architect, those options should help to stimulate new ideas rather than close ideas off.

The Passivhaus approach is pushing higher performance, and there are many other fields of design where very specific performance criteria must be met – in particular laptop computers, for example, where performance and ergonomics are matched with aesthetics to make highly desirable products (this is part of the success of Apple). There is no inherent conflict between performance and design here: in fact, many would argue that there is an interconnectivity between them – limits drive better solutions.

PHPP software

Version 7 of the PHPP, which was released in summer 2012, consists of three spreadsheets and a 217-page supporting manual. The main spreadsheet is complex, consisting of 36 interlinked worksheets. A 'Final Protocol' supplementary

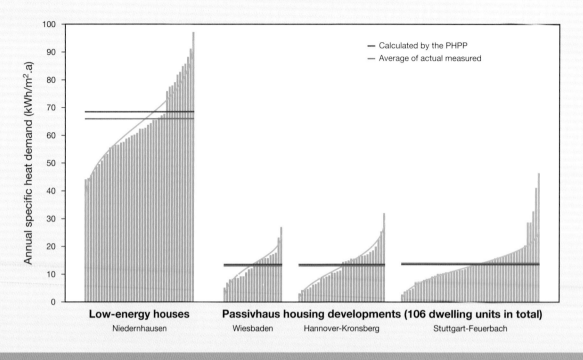

Figure 7.1 Comparison of actual energy performance with that predicted by the PHPP.
Source: Passivhaus Institut

spreadsheet is provided to help with design and commissioning of the heat recovery ventilation system (MVHR), and there is an import/export tool (new in Version 7) that allows the PHPP to link to CAD (computer-aided design) applications. The spreadsheets are designed to run on Microsoft Excel but (although not formally tested) can also be run on Open Office's spreadsheet program. The Passivhaus Institut chose to use the more open format of a spreadsheet rather than develop bespoke software partly because of cost limitations, but also because it is keen to allow users of the PHPP to see how the calculations are derived.

The PHPP can be bought direct from the Passivhaus Institut (www.passiv.de) or (in the UK) from the AECB (www.aecb.net/carbonlite/phpp.php) at a similar price, around £150 (much cheaper than SAP).

The PHPP worksheets

The description in this chapter is not intended to be comprehensive or be a replacement for formal training in using the PHPP, but it highlights the key parts of the model that need to be completed before it starts to produce meaningful information. The screenshots shown here were taken from the previous version of the PHPP; however, the worksheets in Version 7 mostly look very similar. Version 7 also has additional features, including support for a broader range of climates and support for the EnerPHit standard.

The cells in the worksheets are colour-coded to indicate which are for entering data (yellow); which are calculated, show default values, or reference other sheets (white); and which are calculated and display important results (green).

Terms explained

building element – a single material or object comprising part of the structure of a building, i.e. part of a wall, floor or roof.

internal heat gains – the heat gains in a building from its occupants and the use of appliances.

lambda value (λ), also known as k-value – a measure of thermal conductivity. It is measured in W/mK.

psi-value (ψ) – a measure of the rate at which energy passes through a length of material. In a Passivhaus it is used to measure heat loss in a linear thermal bridge. It is measured in W/mK.

specific heat capacity – the amount of heat required to raise the temperature of a unit of a material by a given amount.

thermal bridge – commonly known as a cold bridge – occurs when a material with relatively high conductivity interrupts or penetrates the insulation layer, allowing heat to bypass the insulation.

thermal envelope – the area of floors, walls, windows and roof or ceiling that contains the building's internal warm volume.

treated floor area (TFA) – a convention for measuring usable internal floor area within the thermal envelope of a building.

U-value – a measure of the ease with which a material or building assembly (a structural part of a building made up of a number of building elements) allows heat to pass through it, i.e. how good an insulator it is. The lower the U-value, the better the insulator. The U-value is used to measure how much heat loss there is in a wall, roof, floor or window. It is measured in W/m²K (watts per square metre per degree kelvin [temperature difference between inside and outside the thermal envelope]).

kWh/m².a – kilowatt hours per square metre [of treated floor area (TFA)] per annum.

W/mK – watts per metre per degree kelvin. For linear thermal bridges, this is watts per metre length of the thermal bridge per degree kelvin. (Used to measure psi-value.) For thermal conductivity, this is watts per metre thickness/depth of material per degree kelvin. (Used to measure lambda value.) In both cases the temperature difference measured is that between inside and outside the thermal envelope.

Wh/K per m² – watt hours per kelvin per square metre [of treated floor area (TFA)]. 1 watt hour (Wh) is one thousandth of a kWh (kilowatt hour).

Wh/m³ – watt hours per cubic metre [of air moved]. Used by the PHI to measure the electrical efficiency of mechanical ventilation with heat recovery (MVHR) units.

Verification

The main purpose of the Verification worksheet is to summarise key results of the model. It starts to show meaningful information only when the other important worksheets (described in the following pages) are complete.

Part of the screenshot shown in Figure 7.2 overleaf includes a table called 'Specific Demands with Reference to the Treated Floor Area'; this shows whether the Passivhaus standards are being met. The model assumes an interior temperature of 20°C, and this figure isn't normally changed. However, it is interesting to watch the effect on the building's energy performance of entering a lower or higher internal temperature.

The model has two modes – 'Verification' and 'Design'. In Verification mode, for residential buildings the model calculates the number of occupants by dividing the treated floor area (TFA) by 35m² (the assumed floor area per occupant). In Design mode, if the planned number of occupants is entered, the model uses this figure rather than the derived Verification figure to calculate internal heat gains. In a large house with few occupants, you would use Design mode to check that the ventilation unit is appropriately sized.

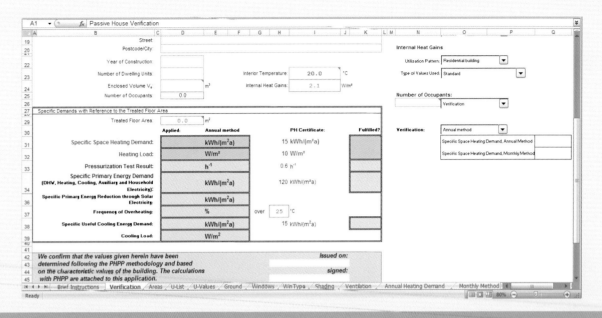

Figure 7.2 Screenshot of the PHPP Verification worksheet.

Areas

Dimensions of all building elements, except the windows, are entered in this worksheet. The information tells the PHPP the external area of the thermal envelope: walls, floor and roof. The use of external dimensions for this calculation in the PHPP is different from the current UK convention, but there is a good, practical reason for this approach, as we will see in the next chapter.

Common mistakes

Note that it is important to ensure that for each line entered in the Area Input section of the worksheet, the correct Group number, e.g. 'Floor slab / basement ceiling' or 'Treated floor area' (Column D) and 'Building element assembly' (Columns T and U) have been selected. Each Building Element Assembly is defined according to the data entered in the U-Values worksheet (see opposite).

Treated floor area (TFA)

As observed in Chapters 1 and 5, Passivhaus uses a stricter definition of what constitutes usable floor area, known as treated floor area (TFA), than that normally applied in the UK. It is very important not to overestimate the TFA in the PHPP, as to do so will result in the PHPP giving over-optimistic figures for the building's energy performance. In summary, the TFA excludes any floor area taken up by:

- external and internal walls
- chimneys and columns with a floor footprint over 0.1m² and over 1.5m high
- stairs with more than three steps
- plant rooms (e.g. hot water storage cupboard), unless plant equipment is wall-mounted, in which case 60 per cent of the floor area of the room can be included. A utility room may or may not be excluded, depending on what plant equipment is in it

- basements, unless within the thermal envelope and without windows, in which case 60 per cent of the floor area can be included
- unheated conservatories (because they are outside the thermal envelope).

It does include:

- floor area within the thermal envelope (i.e. not terraces, balconies, etc.)
- window reveals that are more than 0.13m deep, where the window goes down to the floor
- floor area taken up by fitted shower trays, cupboards and other built-in furniture.

Any floor area where the ceiling height is 1-2m is counted, but only at 50 per cent of the full area; where the ceiling height falls below 1m, the floor area cannot be included in the TFA.

Thermal bridges

The Areas worksheet is also used to enter the length and psi-value of any thermal bridges in the design. In a Passivhaus, the aim is to design out all thermal bridges, but this is not always feasible, especially in retrofits. Where it is not possible, the additional heat losses caused by the thermal bridges need to be entered into the PHPP, to ensure the model remains accurate. The PHPP does not calculate the thermal bridges automatically; this has to be done separately, with the results entered into the worksheet. Chapter 8 looks at thermal bridges and psi-values in more detail.

U-Values

This worksheet calculates the U-value for all building elements except the windows. After you start entering data into the U-Values worksheet, a summary of the U-values is displayed in the U-List worksheet.

In the example pictured below and represented in Figure 7.3 overleaf, an exterior wall built from I-beam and cellulose insulation and a wooden rain screen is described from outside to inside. The rain screen itself is not included in the wall make-up, but does marginally affect the exterior surface resistance (R_{se} – see overleaf). The I-beam is described in Area sections 2 and 3 as percentages of the wall, seen laterally, that are different from the material in Area section 1. Although this method of using percentages is a bit clunky (until you get used to it), it allows the PHPP to model walls with discrete timber structural elements that would otherwise not be possible to model. Where a material used within an external building element has a much higher conductivity (more than four or five times) than the surrounding insulation, the PHPP is not able to model the U-value accurately and additional thermal bridge calculations need to be made.

Two pieces of information are needed for each material within the building element: its thickness (in mm) and its thermal conductivity or lambda value (in W/mK). This value (also known as the k-value) is usually provided by the material's manufacturer; however, the manufacturer's quoted lambda value cannot be taken as read, unless accompanied by a CE mark, which

I-beam wall under construction, represented in the U-Values worksheet example shown overleaf.

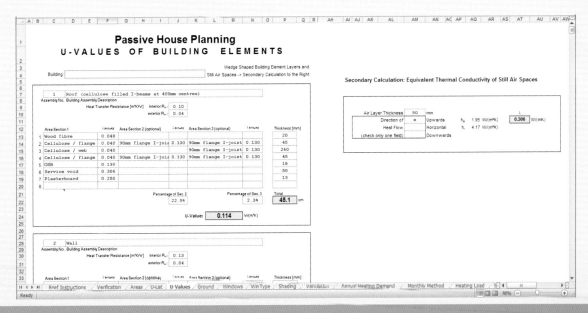

Figure 7.3 Screenshot of the PHPP U-Values worksheet, containing data for the I-beam wall pictured on the previous page.

normally signifies that the conductivity has been determined according to the relevant European Standard (EN) – in this case derived according the Lambda 90/90 convention – see box, right. Without a valid CE mark, a Passivhaus Certifier will require uplift, as much as 20 per cent, in the lambda value. See Appendix B (Thermal conductivity values) for lambda values of some common materials.

The **thermal resistivity** (the material's ability to resist the passage of heat) of the interior surface (R_{si}) and exterior surface (R_{se}) of the building element also need to be entered. In poorly insulated structures, the surface resistivity has an effect on U-values. In ultra-low-energy buildings, even though the U-value is less sensitive to these figures, it is still important to enter R_{si} and R_{se} values in order to get an accurate U-value.

Common mistakes

It is very tempting to use a manufacturer's quoted lambda value, especially if the design's

Lambda 90/90

Lambda 90/90 (λ90/90) values are thermal conductivity values that have been calculated according to the Lambda 90/90 convention. This means that 90 per cent of the test values show a lower conductivity than the stated value, to a statistical confidence level of 90 per cent. Obviously, this refers to materials that are factory-produced and regularly tested. Materials that are blown-in (i.e. 'pumped' in) on-site need density checks, as conductivity is strongly density-dependent. It is much harder to obtain a 90/90 value for materials that are made on-site, such as hemp and lime, especially if there is no quality control of the proportions in the mix.

Lambda 90/90 values are adopted in the UK for Passivhaus calculations.

annual space heat demand is close to or above the 15kWh/m².a limit. But it is important to be conservative in using lambda values, applying the 20-per-cent uplift rule where there is no CE mark.

Ground

The Ground worksheet measures heat losses through the base of the building. If no data is entered into the worksheet, the PHPP will make standard assumptions about these losses, which will be valid for many UK buildings.

WinType

This worksheet contains some technical information that is more fully explained in Chapter 11. As we see in that chapter, windows are a critical element in a Passivhaus or any ultra-low-energy building, and this is reflected in the PHPP. Because of the way the PHPP works, it is more practical to complete the WinType worksheet before the Windows worksheet (see page 100). The WinType worksheet is divided into two halves. In the first, information is entered about the glazing units; in the second, information is entered about the window frames. Figure 7.4 overleaf shows the two sections of the worksheet.

Both sections give values for some standard and generic window glazing and frame types, as well as for examples of Passivhaus-certified components. The rows in yellow at the beginning of each section allow Passivhaus Designers/Consultants to enter details of the windows they are planning to use if these are not listed.

All the information entered into this worksheet and the Windows worksheet allows the PHPP to generate a whole-window U-value (U_w) and an installed whole-window U-value ($U_{w,\,installed}$) for each window. This contrasts with the whole-window U-value quoted by manufacturers, which is based on a window of standard size and configuration (see box on page 172, Chapter 11). Window manufacturers often seem reluctant or unable to provide all the energy performance information needed for the WinType worksheet. However, manufacturers of Passivhaus-suitable and Passivhaus-certified windows should be able to do so.

Glazing information

Two pieces of information are needed:

- g-value – measure of the percentage of energy from the sun that passes through the glazed unit to the inside. The g-value is normally expressed as a fraction between 0 and 1.
- U_g-value – the U-value of the glazed unit (as distinct from the whole window), taken through the centre of the pane, measured in W/m²K.

g-values

The g-value is one of the critical factors in determining the amount of solar gain window glazing of any given size will deliver. In a cool-temperate climate, where winter solar gain is desirable, windows on any building façades that are subject to solar gain should have a g-value of above 0.5; ideally 0.6 or more. In a hot climate, where cooling load is more of an issue than heating load, lower g-values, below 0.35, are preferable. Table 7.1 on page 99 shows typical g-values for different types of glazed units.

U_g-values

A Passivhaus-certified window requires a U_g of less than 0.75W/m²K, for a vertically inclined window (i.e. not a roof light) in a cool-temperate climate[1] such as the UK's, and a triple-glazed, argon-filled unit with two 14mm or bigger spacers will comfortably achieve this. U_g requirements for inclined and horizontally mounted roof lights are a little less strict: 1.00 and 1.10W/m²K respectively. See Chapter 11 (pages 170-6) for more information on window U-values.

Frame information

The characteristics of window frames have a considerable impact on the overall performance of the window. The pieces of information required in the PHPP are:

- U_f-value – the U-value of the frame (measured in W/m²K).
- Width of each frame (top, left, right, bottom), measured in metres.
- Psi-value (ψ) of the spacer (ψ_{spacer}).
- Psi-value (ψ) of the junction where the frame meets the wall ($\psi_{installation}$).

Frame U-value (U_f-value)

While manufacturers find it relatively easy to make triple-glazed units with a low enough U_g-value for a Passivhaus, to date the frames have been more challenging. A Passivhaus-certified window should have a U_f-value of below 0.80W/m²K. A few manufacturers have achieved this, but many have not. Either way, it is good practice to specify windows where the frames comprise not too high a percentage of the total window area. Even at a U-value of 0.80, windows perform poorly relative to walls in an ultra-low-energy home. Whereas the glazed area

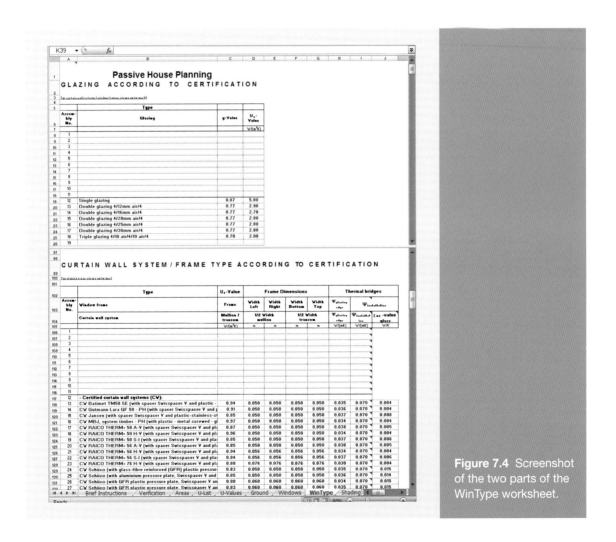

Figure 7.4 Screenshot of the two parts of the WinType worksheet.

performs the key functions of providing daylight and allowing in solar energy, the frames, of course, do neither.

Psi-value of spacer

Like the frame, the spacer is an energy-performance weak point in windows, and its psi-value has a significant impact on the window's overall performance. 'Warm-edge' or insulated spacers have psi-values below 0.05W/mK; the best ones below 0.03W/mK.

Psi-value of frame–wall junction

The installation psi-value (frame–wall junction) is also a significant determiner of the installed window's overall performance. In the PHPP, the

Table 7.1 **Typical g-values for different glazed unit types**

Glazing type	Suitable for a Passivhaus	Typical g-value
Single-glazed unit	No	0.87
Air-filled double-glazed units	No	0.77
Air-filled triple-glazed units	Yes*	0.70
Argon-filled double-glazed units	No†	0.56-0.64
Argon-filled triple-glazed units (standard float glass)	Yes	0.50
Argon-filled triple-glazed units (low-iron glass)	Yes	0.60

* If the U_g value is below 0.75W/m²K.
† But can be used in Passivhaus buildings in warmer climates.[2]

Argon-filled versus krypton- or xenon-filled glazed units

Triple-glazed, argon-filled units are the ones used in most Passivhaus buildings located in cool-temperate climates. Krypton or xenon fillings are also options but are a lot more expensive, not least because these are much rarer gases than argon. Also, krypton- or xenon-filled windows usually have narrower spacers (typically 10mm or 12mm) than argon- or air-filled units (which tend to have spacers of 18mm, sometimes 20mm, although older or very large units have slimmer spacers). Argon, krypton and xenon gradually leak from the glazed units and most of the gas can be lost within ten years. For a krypton- or xenon-filled unit with narrow spacers, its performance will be much more degraded for a good proportion of its life (after the gas has leaked out) than the performance of an argon unit with 18-20mm spacers (after the gas has leaked out). With spacers of this size, even after the argon has leaked out, the window should perform sufficiently well to be suitable for a Passivhaus. From an economic and environmental perspective, use of krypton or xenon is therefore generally to be discouraged. The PHPP will help the designer avoid their use.

designer should assume an installation psi-value of 0.04W/mK, unless separate thermal bridge modelling can demonstrate a lower value. Chapter 11 covers the principles of how to detail window–wall junctions to minimise thermal bridging and air leakage.

Common problems

The window technical data described above has a surprisingly large impact on the building's performance. Getting accurate figures from the window manufacturer is therefore very important. Manufacturers who fabricate windows for Passivhaus buildings should be able to provide this data. However, in a project where the design is to a less demanding fabric energy standard, such as the Code for Sustainable Homes (CSH) or the **AECB Silver Standard**, poorer-performing windows can be used, in which case it is likely that there will be no frame U-value (U_f) or spacer psi-value (ψ_{spacer}) available. These values could be individually calculated in THERM (see Chapter 8); otherwise very conservative values would need to be entered.

Windows

The Windows worksheet is used to compile information about each window in the building. Most of this can be taken directly from the building's window schedule (a list of all the windows in a building), except when a window consists of multiple casements, in which case each casement must be entered as a separate window. The following information needs to be entered for each window:

- A short description of the window (usually taken from the window schedule).
- Deviation from north – tells the PHPP what direction the window faces.
- Angle of inclination from horizontal – normally 90°, except for roof lights.
- Width and height of the structural or 'rough' opening. If the window is round, enter dimensions for a square of equivalent area. This will give a more conservative result, as a square window will have more frame than a round or oval one of equal area.
- A façade into which the window is being installed (the façades are defined in the Areas worksheet).
- The window's glazing (choose a glazing type that has been defined in the WinType worksheet).
- The window's frame (choose a frame type that has been defined in the WinType worksheet).
- Installation – for each window side, enter '1' where a window side is adjacent to a wall or '0' where it is adjacent to another window. This determines whether or not the heat losses due to the installation thermal bridge are applied. Where a window is partly bounded by another window and partly by a wall, enter a fraction to reflect this.

Mullions

There is also a facility in this worksheet to describe windows with mullions. However, as frames (and mullions) and the spacers that they necessitate are energy-performance weak points in a window, specifying windows subdivided into mullions will significantly increase their heat losses and decrease the solar energy captured through the glazing. This is an instance where Passivhaus can discourage certain designs on costs grounds (though not necessarily rule them out). Currently, Passivhaus-suitable windows are mostly made on the Continent or copy Continental window styles, where mullions are uncommon in new builds. As window manufacturers in each country start to respond to growing demand for ultra-low-energy windows, they will design ones that reflect domestic architectural styles. Windows with mullions will always carry an energy-performance penalty, but this does not mean they cannot be used in a Passivhaus, except perhaps in extremely cold climates.

The left and right (and centre) sections of mullioned windows need to be entered in the WinType worksheet as separate window types (see page 97), owing to the different frame widths in each mullioned section.

Curved windows

If the design includes a curved window as part of a curved façade, enter the 'deviation from north' (direction of the window) of the lateral midpoint of the window and measure the dimensions as if it were flat, as shown in Figure 7.5 below.

Southern hemisphere

If the PHPP model is for a building in the southern hemisphere, the window 'deviation from north' figures need to be inverted. For example, if the property being modelled is in Australia, and has a north-facing window, the 'deviation from north' should be entered as '180', not '0' as would be the case if it was in the northern hemisphere.

Common mistakes

Unless the design is for a house on a straight north–south or east–west axis, it can be quite easy to enter the wrong information in the 'deviation from north' column. Likewise, you should double-check the 'window rough openings'. It is also easy to overlook the need to enter each section of a mullioned window as a separate entry.

Shading

As noted in earlier chapters, shading is important in Passivhaus and ultra-low-energy building design. Winter shading will reduce desirable solar gain and add to the building's heating requirements and is therefore to be minimised where possible. Summer shading produces unwanted solar gain, resulting in a greater risk of overheating. The Shading worksheet deals only with winter shading. It picks up information that has been entered into the WinType and Windows worksheets, then calculates the glazed area based on the window opening sizes and frame dimensions provided in the Windows and WinType worksheets respectively. Shading information needs to be added for each window. The information required in this worksheet is as follows (all items, except the last one, are measured in metres).

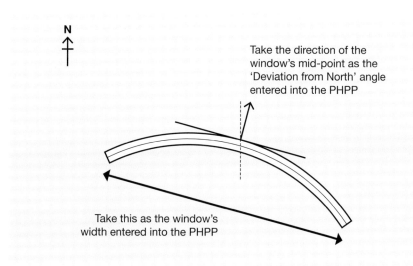

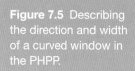

Figure 7.5 Describing the direction and width of a curved window in the PHPP.

- Height of shading object (h_{hori}) – taken from the top of the shading object to the base of the glazed area of the window being shaded (see Figure 7.6 below).
- Horizontal distance to shading object (d_{hori}) – taken from the top of the shading object to the external glazed surface (see Figure 7.6 below).
- Window reveal depth (o_{reveal}) – the distance between the external surface of the glazing and the external surface of the wall (see Figure 7.7 opposite).
- Distance from glazing edge to wall reveal (d_{reveal}) – in a window without mullions, this is a simple figure. In a mullioned window, it is a bit more cumbersome: the average (arithmetic mean) of the two d_{reveal} figures must be taken. In the example in Figure 7.7, the average (mean) of $d(left)_{reveal}$ and $d(right)_{reveal}$ would be entered into the PHPP. In a window with many mullions, this calculation would need to be repeated for each mullioned section – rather time-consuming!
- Overhang depth (o_{over} – see Figure 7.8 opposite).
- Distance from upper glazing edge to overhang (d_{over} – see Figure 7.8).

Additional shading reduction factor (r_{other}) – this is entered as a percentage, where (rather counter-intuitively) 100% means no additional shading and 0% means total shading. If it is left blank, the PHPP assumes that there is no additional shading. You should enter a figure only if there is a tree or some similar object that partly shades the window. A useful visual representation of any additional shading can be gained by loading the architectural site and building plans into Google SketchUp (www.sketchup.com) or similar and setting the site's longitude and latitude; this facilitates a more accurate estimate of the shading percentage.

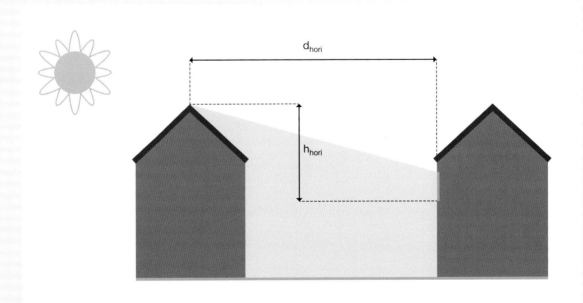

Figure 7.6 h_{hori} and d_{hori} in the Shading worksheet.

Chapter Seven • Using the Passivhaus Planning Package (PHPP)   103

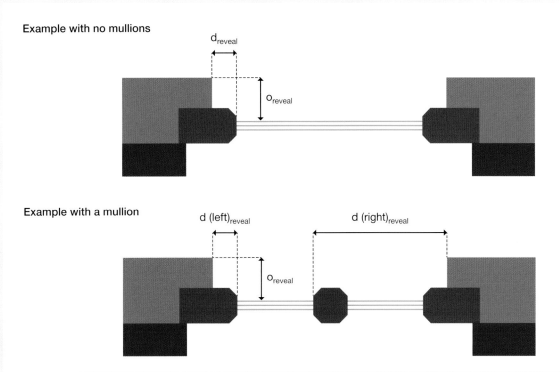

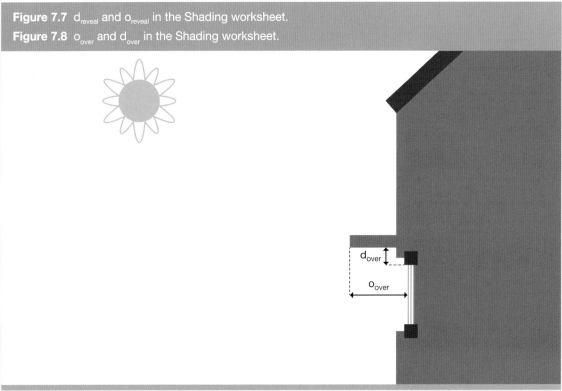

Figure 7.7 d_{reveal} and o_{reveal} in the Shading worksheet.
Figure 7.8 o_{over} and d_{over} in the Shading worksheet.

Ventilation

The Ventilation worksheet is divided into two parts. The first, shown in Figure 7.9 below, is used to compile information to determine the size of the ventilation system required. In a Passivhaus, the MVHR system supplies fresh air to living and sleeping areas, and extracts stale air from 'wet-room' areas (kitchens, bathrooms and WCs). It also recovers heat from the extracted air when needed (during cold weather) and supplies it to the new air. The number of kitchens, bathrooms, showers and WCs is entered into the worksheet, and an airflow rate of 30m³/hr per person is recommended and is assumed by default, as are recommended extraction rates for different extraction points.

The number of air extraction points may need to be changed or entered. The PHPP calculates an average airflow rate from this information.

(The supplementary Final Protocol spreadsheet, provided with the PHPP, allows you to design the MVHR system in more detail. Figures from this spreadsheet need to be entered manually into the Ventilation worksheet of the PHPP. The Final Protocol spreadsheet can also be used to commission an MHVR unit.)

Next, two 'wind coefficient' factors should be entered to describe the degree to which the building is protected from wind. In this case, approximate data is adequate for the purposes of the PHPP. Below this there is a box to enter

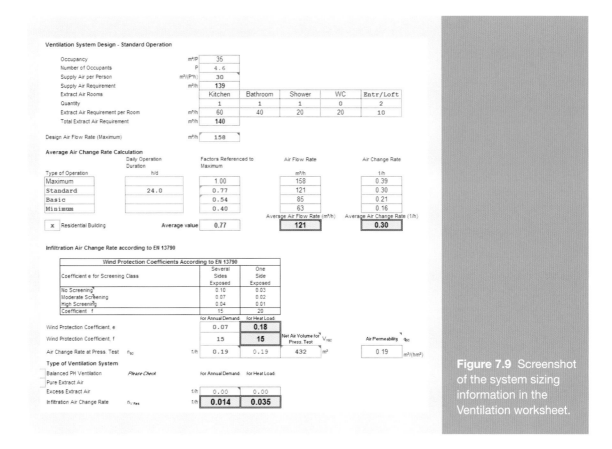

Figure 7.9 Screenshot of the system sizing information in the Ventilation worksheet.

Chapter Seven • Using the Passivhaus Planning Package (PHPP)

the measured air change rate through the building fabric, taken from the building's first or most recent airtightness test, during which the building is pressurised and depressurised by 50Pa (pascals) above and below ambient atmospheric pressure (see Chapter 9). The PHPP assumes the maximum allowable Passivhaus airtightness value of 0.6ach until the results of the air test are entered.

The section of the Ventilation worksheet below this (see Figure 7.10 below) is used to describe the MVHR system. First, indicate whether the MVHR unit is to be located inside or outside the thermal envelope. Second, select the MVHR unit using the drop-down box. If the unit to be used is not in the list, add one in the 'User defined' section.

Less common than a standard MVHR unit, a compact unit may be used (this combines MVHR, hot water and heating in a single 'box'). Its MVHR specification will be displayed automatically in the Ventilation worksheet once the unit's details have been entered into the PHPP's Compact worksheet. Compact units are likely to become more popular in coming years, as more achieve Passivhaus-certified status. Their chief benefit is space saving – which is an important consideration.

Requirements for a Passivhaus heat recovery (MVHR) unit

As we saw in Chapter 3, the MVHR unit is one component of a Passivhaus that should be

Figure 7.10 Entering information on the MVHR unit and the ducting between the unit and the thermal envelope in the Ventilation worksheet.

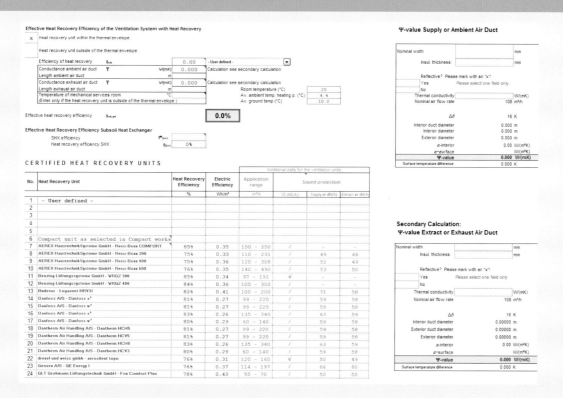

certified if the building is to meet the Passivhaus standard. The worksheet requires six values for the MVHR unit, as shown in Table 7.2 below, along with the criteria that these must meet for a Certified Passivhaus MVHR unit.

The lengths and psi-values of the sections of ducting between the MVHR unit and the boundary of the thermal envelope also need to be entered. These are critical to the efficiency of the MVHR system as a whole. If the MVHR unit is located within the thermal envelope, these ducts are the intake (or 'ambient') and exhaust ducts carrying cold air. If the MVHR is located outside the thermal envelope, these ducts are the length of supply and extract ducts carrying warm air between the MVHR and the boundary of the thermal envelope (see Chapter 12, page 200). The psi-values of the ducts have to be calculated in a pair of 'Secondary Calculation' boxes. For each of these secondary calculations, the following information is needed:

- Diameter of the duct (in mm) – in a single residential dwelling this is normally 160-180mm.
- Thickness of the insulation (in mm) – minimum 50mm.
- Whether the external surface is covered in reflective material.
- The conductivity (lambda value) of the insulation around the duct.

The insulation material must be impervious to water, so a material such as Armaflex® needs to be used. As is discussed in Chapter 12, the ducts between the MVHR unit and the thermal envelope should be kept as short as possible. If they are overly long or inadequately insulated, this will have a significant effect on the building's annual heat demand. Tweaking these variables in the PHPP illustrates this point well.

Table 7.2 **MVHR values**

Variable	Passivhaus criterion
MVHR unit heat recovery efficiency	75%* or better
MVHR unit electrical efficiency	Maximum 0.45Wh/m³ [of air moved]
MVHR unit application range: the unit's designed range of airflow rates (m³/hr). The capacity of the unit should be approx. one-third greater than the maximum 'Design Air Flow Rate' calculated as being required for the project in the Ventilation worksheet (see Figure 7.9 on page 104).	
Sound emitted by the MVHR unit (either in plant room or adjacent usable room)	Maximum 35dB(A)†
Sound emitted in the supplied rooms	Maximum 25dB(A)†
Sound emitted in the extracted rooms	Maximum 30dB(A)†

* As determined independently using Passivhaus Institut methodology
† 'A-weighted' decibels

Annual Heating Demand

As the other key worksheets are completed, this worksheet starts to provide a useful summary of the building's heat losses and gains. The assumed (default) room height of 2.5m (V_v) can be changed.

Summer

The Summer worksheet calculates the percentage of summertime overheating. The Passivhaus standard includes a limit (10 per cent) on the number of hours per year when the internal temperature exceeds 25°C. Some of the building-specific factors that affect this, and need to be entered or checked (i.e. are the default values correct for your build?), are given below.

- Specific heat capacity – this is a measure of the building's thermal mass. There are three suggested values (60, 132 and 204Wh/K per m^2 of treated floor area [TFA]) – another example of where the PHPP requires only an approximate value in order to model the building's performance accurately. The default value is 60, which assumes a low thermal mass.
- Overheating limit: 25°C by default, in line with the Passivhaus standard. It is possible to change this value in order to get a broader view of the extent of overheating, which is useful.
- Air change rate from natural ventilation and/or mechanical ventilation.
- Additional summer ventilation for cooling – indicate whether this is to be achieved by manual opening of windows at night-time or by an automated system.

The resulting frequency of overheating percentage is then displayed. Whether 10 per cent overheating, although allowed, is acceptable is for you to decide. We suggest a much lower figure, especially in view of uncertain future climate patterns.

Shading-S

The Shading-S worksheet picks up data from the Windows and Shading worksheet and provides two extra columns to indicate what additional shading is provided in summer. The temporary shading reduction factor (z) describes shading from blinds, awnings or other shading devices. As with the entry of the additional shading reduction factor (r_{other}) in the Shading worksheet, using Google Sketchup helps to make a more accurate estimate of the percentage shading. Where the shading device will be operated manually, the PHPP convention is to apply a factor (70 per cent) to model the assumption that we don't always remember to operate the shade during hot weather. Shading is based on external devices, as this is the only truly effective shading option; you can enter values for internal shading devices, but these will have a much more limited effect.

DHW + Distribution

The Domestic Hot Water + Distribution worksheet calculates losses through the building's hot water pipes. Pipes are defined as either 'circulation' or 'individual' pipes. This worksheet models the heat losses of both or either type of pipework.

Circulation pipes are pipes that form a hot water circuit to and from the hot water store. A low-power pump continuously pushes hot water around the circuit, effectively making the circuit an extension of the hot water store. Circulation pipes are used in larger buildings, such as hotels, so that hot water can be made available quickly over a much wider area without having to draw off and waste large quantities of hot water every time a tap is turned on. Circulation pipes have an energy cost, owing to both the heat losses in the pipework and the electricity needed to run the circulation pump.

Individual pipes are pipe runs to end points of use (taps, shower heads, washing machines, etc.), either direct from the hot water store or from the hot water circuit, if there is one.

Wherever possible, in standard-sized residential units or homes, it makes sense to place the hot water store centrally, midway between the points of use, so that the need for a hot water circuit can be avoided. Using a narrower bore of pipework for individual pipes also helps to reduce energy use, because a smaller volume of water is contained within the pipework. Clearly, a balance has to be set between energy saving and providing a reasonable flow rate.

The PHPP assumes that any heat in the individual pipes will be lost, as it calculates on the basis that each tap or point of use will be used three times a day – enough time between uses for the heat to dissipate, even if the pipes are insulated.

The worksheet assumes that each person will use 25 litres of hot water daily. This is lower than Standard Assessment Procedure (SAP) but the default figure in cell K34 can be changed if needed; the PHPP suggests 12 litres for office accommodation.

SolarDHW

The SolarDHW worksheet models the contribution of a solar thermal hot water system, should the building have one. The worksheet takes as its starting point the net demand for domestic hot water as calculated in the DHW + Distribution worksheet, and the site's location from the Climate worksheet (see Figure 7.11 opposite).

The following information is needed to describe a solar hot water system:

- Type of solar collector – in most cases, this will either be an 'improved flat plate' or a 'vacuum tube' collector.
- Solar collector area – this can be taken from the manufacturer's specification; ensure that the area is the net collector area, not the total collector footprint.
- Deviation from north – enter 180° if the collectors face due south. In the southern hemisphere, enter 180° if the collectors face due north.
- Inclination – i.e. roof pitch.
- Any shading – for example, if the site is in a steep valley or is partly shaded by trees. There are three fields for this.
- Type of solar storage – select from a range of store types and sizes or specify a bespoke type, if sufficient corroborated technical information is available from the manufacturer.

Although solar hot water is not a requirement of Passivhaus, its incorporation in the design will significantly reduce the building's energy use, assuming the site, location and orientation are suitable, and is therefore encouraged by the PHI (hence the worksheet).

Climate

The Climate worksheet is a key component of the PHPP, as accurate climate data is vital to creating an accurate PHPP model. Monthly temperature and solar radiation data is required. By default, the Climate worksheet is set to a standard central European climate model. It is essential to change this to a climate model relating to the location of the project.

The PHPP contains partial and full data sets for selected locations around Europe and North America on a drop-down menu. The complete

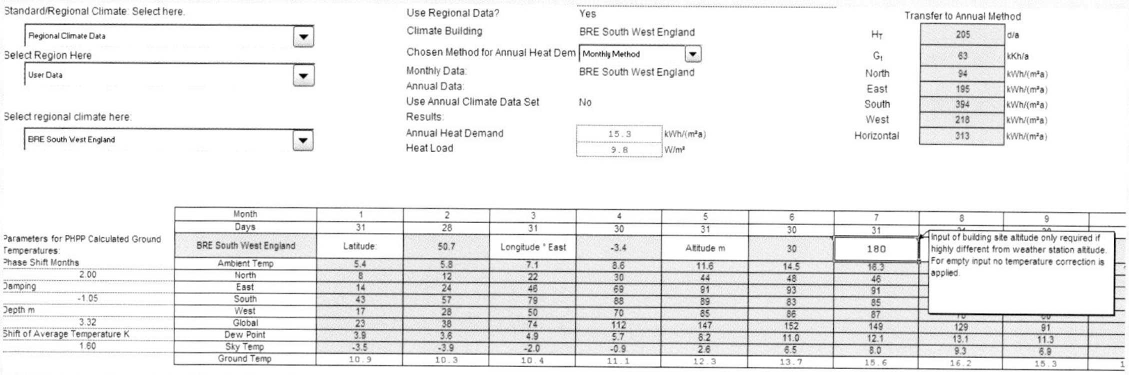

Standard/Regional Climate: Select here.
Regional Climate Data
Select Region Here
User Data
Select regional climate here:
BRE South West England

Use Regional Data? Yes
Climate Building BRE South West England
Chosen Method for Annual Heat Dem Monthly Method
Monthly Data: BRE South West England
Annual Data:
Use Annual Climate Data Set No
Results:
Annual Heat Demand 15.3 kWh/(m²a)
Heat Load 9.8 W/m²

Transfer to Annual Method

H_T	205	d/a
G_t	63	kKh/a
North	94	kWh/(m²a)
East	195	kWh/(m²a)
South	394	kWh/(m²a)
West	218	kWh/(m²a)
Horizontal	313	kWh/(m²a)

Parameters for PHPP Calculated Ground Temperatures:

Phase Shift Months	2.00
Damping	-1.05
Depth m	3.32
Shift of Average Temperature K	1.60

Month	1	2	3	4	5	6	7	8	9
Days	31	28	31	30	31	30	31		
BRE South West England	Latitude:	50.7	Longitude ° East	-3.4	Altitude m	30	180		
Ambient Temp	5.4	5.8	7.1	8.6	11.6	14.5	16.3		
North	8	12	22	30	44	48	46		
East	14	24	46	69	91	93	91		
South	43	57	79	88	89	83	85		
West	17	28	50	70	85	86	87		
Global	23	38	74	112	147	152	149	129	91
Dew Point	3.9	3.6	4.9	5.7	8.2	11.0	12.1	13.1	11.3
Sky Temp	-3.5	-3.9	-2.0	-0.9	2.6	6.5	8.0	9.3	6.9
Ground Temp	10.9	10.3	10.4	11.1	12.3	13.7	15.6	16.2	15.3

Figure 7.11 Amending the default altitude in the Climate worksheet.

data sets allow the PHPP to calculate both the annual space heat (or cooling) demand and the heat (or cooling) load; the partial data sets only allow the PHPP to calculate annual heating (cooling) demand.

If the project being modelled is located in a climate that is very different from any of the available climates in the PHPP, there are two options:

- If the site is at high altitude, amend the default altitude for the chosen climate data set. In Figure 7.11 above, the default altitude of 30m has been changed to 180m. Change the default only if the difference in altitude is significant (say, 50m or more).
- If the site has a microclimate that deviates greatly from the nearest climate model available in the PHPP, obtain climate data for the project's specific location. BRE (www.bre.co.uk) and Meteonorm (www.meteonorm.com) both offer site-specific climate data for the PHPP. In the UK, BRE has created 22 geographically broader regional data models that have been ratified by the PHI and are currently free to download from www.passivhaus.org.uk/regional-climate-data.jsp?id=38. Version 7 (2012) of the PHPP already includes this data.

RECAP

The PHPP incorporates a huge body of knowledge, and, although from the outline given in this chapter it may seem daunting, after the initial effort to familiarise yourself with it you will find that it is an exciting, educational and cost-effective tool. It provides assurance that your project is on target, which in itself is very valuable. While it is not the only input into a successful design, it is one that is inadvisable to avoid. Quantified facts are a necessary part of the mix to ensure that properly informed decisions are made and, crucially, that the design's intended energy performance is achieved.

CHAPTER EIGHT
Thermal bridges

Constructional and geometrical thermal bridges, linear and point thermal bridges, thermal bypass, internal and external psi-values, dealing with thermal bridges, thermal bridge calculation

The subject of thermal bridges is, of necessity, technically complex. It is therefore important to ensure that your professional team has sufficient understanding of thermal bridges and how to avoid them, in order to ensure that the design eliminates them (or, in retrofits, minimises them).

In buildings, thermal bridges (commonly known as 'cold bridges'), occur when a material with relatively high conductivity interrupts or penetrates the insulation layer. A thermal bridge provides a 'path of least resistance', allowing heat to bypass the insulation and significantly reducing its performance.

There are three reasons to avoid thermal bridges:

- In poorly insulated buildings, the additional impact of thermal bridges is, as a percentage of the total heat loss, relatively small. As insulation levels are increased, the relative impact of thermal bridges grows dramatically, such that it would not be possible to reach the Passivhaus standard without effectively addressing them. In a retrofit aiming for the EnerPHit standard, thermal bridges need particular attention: although they cannot generally be designed out, they can be minimised.
- Those sections of the thermal envelope where thermal bridging occurs will have lower internal surface temperatures in cold weather, resulting in increased relative humidity (RH – the quantity of water vapour in a given volume of air) on the surface, which in turn increases the risk of mould or condensation. This is a common phenomenon in poorly insulated buildings and one where the causes are often incorrectly diagnosed.
- Thermal bridges are difficult and time-consuming for the Passivhaus Designer to calculate (as we saw in Chapter 7, the PHPP does not do this automatically). Designing them out makes it much simpler (and more cost-effective) to create an accurate energy model of a building in the PHPP. Where a thermal bridge exists and no separate thermal bridge calculation has been made, the Passivhaus Certifier will make very conservative assumptions.

When do thermal bridges arise?

Thermal bridges can form wherever there is a junction between different building components or where there are corners in the thermal envelope (usually the external walls). The different types of thermal bridge are described below.

Constructional thermal bridges

Inexperienced Passivhaus designers may think that there are insurmountable conflicts between the need to build a structurally sound building and an energy-efficient one. It is true that materials that are commonly used for their structural properties, such as steel, concrete or even wood, are not generally good insulators, and in conventional structures they often penetrate insulation at building junctions, causing 'constructional thermal bridges'.

The image overleaf, from a 1970s house before retrofitting work, shows a reinforced concrete lintel above a veranda door. The surface temperature (represented by colour in the thermographic picture) of the lintel is similar to that of the single-glazed window frame to its right, and is a good example of a thermal bridge. Other examples of constructional thermal bridges are window-to-wall joints, wall ties in a filled cavity wall, and timber in an insulated timber-frame construction.

Example of a constructional thermal bridge: a reinforced concrete lintel above a veranda door.
Image: Infrared Thermal Imaging Surveys UK

Figure 8.1 opposite shows an architectural detail and thermographic images of a wall-to-floor junction where a mistake with the detailing has left a gap in the insulation, causing a constructional thermal bridge.

In a building based on structural timber, it is very easy to create thermal bridging unintentionally. It is essential, therefore, to plan the layout of the studwork so that such effects are minimised while ensuring that structural requirements are still met. A good example of where this was not done can be seen in the photograph opposite, showing an unnecessary clustering of studwork. In particular, careless use of prefabricated elements, such as structural insulated panels (SIPs), will create unnecessary thermal bridging.

Thermal bridges arising from a building's geometry

The corners of building elements cause thermal bridging simply because of their shape, unless

Thermal bypass

Thermal bypass can sometimes be confused with other types of thermal bridge, which is why it is mentioned here. It is different from the other effects described in this chapter. The main practical point is that insulation must always be installed without gaps!

Air can be a good insulator when it is static in a sealed space, if the space is narrow enough and if it is not perpendicular to the thermal envelope (e.g. in the sealed space of a double-glazed window). However, as soon as there is any air movement it becomes an effective carrier of heat, which can seriously compromise the performance of an insulation layer. Common examples include an unfilled cavity in a typical UK cavity wall or in poorly laid loft insulation. Thermal bypass occurs where heat is carried by convection or by 'blow through': the former is where air circulates within a space, the latter is where air passes through the space from the inside to the outside of the thermal envelope. Thermal bypass can reduce the expected performance of an insulated building element by 40-70 per cent.[1]

Chapter Eight • Thermal bridges 113

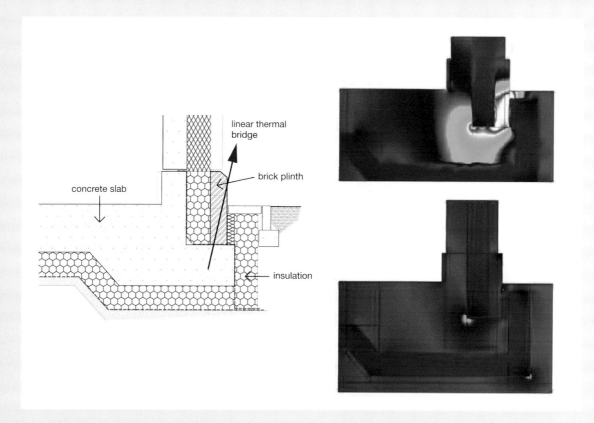

Figure 8.1 An example of a thermal bridge on a wall-to-floor junction where a gap in the insulation is created by a brick plinth, and infrared modelling of before and after the remedy. (The remedy is likely to have been insertion of a load-bearing insulating block such as Foamglas® Perinsul.)

they are intentionally designed to avoid them (as we see on pages 116-17, using the external psi-value convention, a junction is deemed to be 'thermal–bridge-free' if its external psi-value is less than 0.01W/mK). Figure 8.2 overleaf shows a solution that meets the structural requirements of the building – in this case a single-storey structure. The dotted black line represents a flexible lining that abuts the neighbouring building and contains the cellulose insulation infill. It allows the I-beams to be recessed by 50mm. This, plus the positioning and limited number of the I-beams at the corner, eliminates thermal bridging at the building junction. Examples of geometrical thermal bridging all

An example of poor timber framing, where the effect of over-designing the structural frame creates multiple thermal bridges.

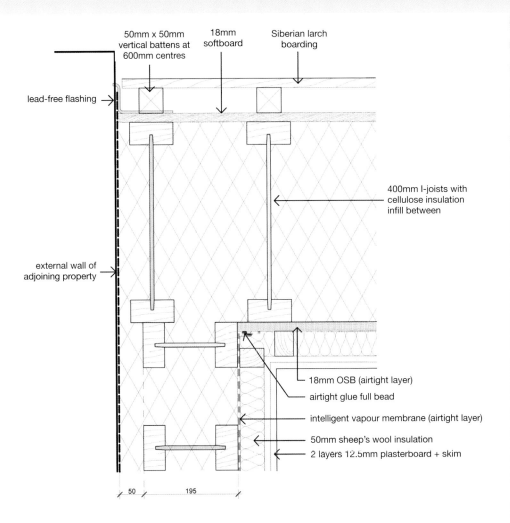

Figure 8.2 Architectural detail of a corner junction of two walls that avoids thermal bridging. The photograph on the right shows how it was executed on-site. This detail relies on blown-in insulation – so consider where the access hole for the pump will be for each section of the wall.

occur in situations where the external heat loss area is different (usually larger) than the internal heat loss area: in a wall-to-floor junction, a wall-to-roof junction, or the vertical corner joints between walls. Most geometrical thermal bridges have an element of constructional thermal bridging too.

Measuring thermal bridges

This section looks at the units of measure used to quantify a thermal bridge, as well as at the conventions for measuring them – both are important in understanding thermal bridges. The actual calculation of thermal bridges is referred to at the end of this chapter.

Point thermal bridges

Some construction details cause point thermal bridges, usually occurring in repeating patterns: for example, mechanical fixings through insulation, or structural connection points at a junction. The photo below shows thermally

Terms explained

chi-value (χ) – the rate at which heat passes through a material that penetrates another material at a point, where the penetrating material conducts heat better than the surrounding material; measured in **W/K**. Used to measure heat loss in a point thermal bridge.

psi-value (ψ) – the rate at which energy passes through a length of material, measured in **W/mK**. Used to measure heat loss in a linear thermal bridge.

W/mK – watts per metre per degree kelvin. For linear thermal bridges, this is watts per metre length of the thermal bridge per degree kelvin temperature difference between inside and outside the thermal envelope. (W/mK is also used to measure thermal conductivity / lambda value.)

W/K – watts per degree kelvin [temperature difference between inside and outside the thermal envelope].

broken fixings, i.e. made to minimise point thermal bridging, to attach external wall insulation.

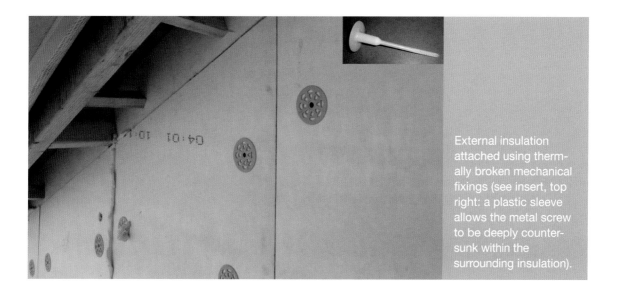

External insulation attached using thermally broken mechanical fixings (see insert, top right: a plastic sleeve allows the metal screw to be deeply countersunk within the surrounding insulation).

The heat loss of a point thermal bridge is described by its **chi-value** (χ). Point thermal bridges are deemed by the PHI to be 'thermal-bridge-free' (i.e. insignificant – see overleaf) unless they are caused by steel or other penetrations with good conductivity. Where there are recurring steel thermal bridges, such as traditional metal cavity wall ties, additional thermal bridge modelling would be required and the results entered into the PHPP.

Point thermal bridges are not normally created purely as a result of geometry. One exception would be in a cone-shaped building – not a common building form!

Linear thermal bridges

Linear thermal bridges can be caused by construction details or may be formed as a result of a building's geometry. They often occur where a length of one building element or material meets another – for example, where a wall meets a floor, roof or window. They can also exist within a wall or other building element – for example, where poor installation has left a gap between the edges of two pieces of insulation.

The former can be avoided by careful design of the building junction (see page 118); the latter by taking sufficient care to fit the insulation accurately on-site. When using board-based insulation, this means very careful cutting to ensure less than 3mm gaps. Softer roll-based and blown-in insulation requires less skill and time to achieve accurate installation. The building designer should take this into account when specifying.

The specific heat loss (i.e. the heat loss for a given length) of a linear thermal bridge is described by its psi-value (ψ).

'Thermal-bridge-free'

As a rule of thumb, a junction in a Passivhaus is considered to be thermal-bridge-free if, on examination of the construction, the insulation can be seen to be continuous at a minimum of two-thirds the thickness of the insulation surrounding the junction (assuming all the insulation has the same thermal conductivity). Quantified, a junction can be declared thermal-bridge-free if the external psi-value (ψ_e) is 0.01W/mK or lower.

Internal versus external psi-values

There are two conventions for measuring a linear thermal bridge. They produce different psi-values in thermal bridges caused by geometry, e.g. of the type found in junctions between building elements, illustrated in the simplified wall-to-floor junction opposite (Figure 8.3). The general UK convention (outside Passivhaus) is to use internal dimensions to measure the thermal envelope and calculate heat loss, based on U-values multiplied by areas. In this case, the psi-value should also be calculated on the same assumption. Thus, internal dimensions of the junction are taken to calculate an internal psi-value (ψ_i). In Passivhaus, the convention is to use the external dimensions of the junction to calculate an external psi-value (ψ_e), which is why it is important to enter the external dimensions of the thermal envelope in the PHPP.

In all geometry thermal bridges, such as the one illustrated in Figure 8.3, there are heat losses associated with the two elements (in this example, the wall and the floor) and also the junction (in this case the corner – shown shaded in orange); these are measured using their respective U-values. In the example illustrated, the extra heat loss associated with the thermal bridge relates to the corner.

Chapter Eight • Thermal bridges 117

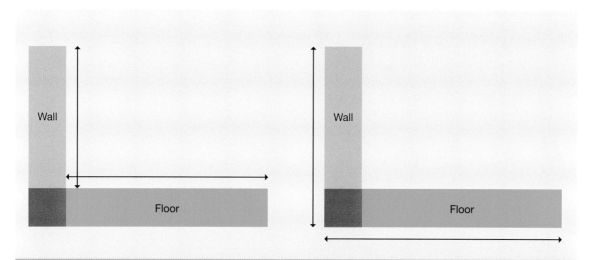

Figure 8.3 Simplified drawing of a wall-to-floor junction, showing (left) internal dimensions used to calculate an internal psi-value (ψ_i) and (right) external dimensions used to calculate an external psi-value (ψ_e).

The difference between measuring the external dimensions and internal dimensions is critical, because the former includes the heat loss relating to the corner and the latter excludes it.

If the corner junction contains only the same materials as the wall or the floor, its external psi-value (ψ_e) will be a negative figure. This negative value can be entered into the PHPP and will trim the annual heat demand; this is useful only if you are looking for savings on a project that has just missed the 15kWh/m².a limit and where other options are too impractical or expensive.

As we see later in this chapter (page 120), once the heat loss factors described above are known, a separate calculation is needed to convert these into either internal (ψ_i) or external (ψ_e) psi-values.

The above may seem a rather laboured and perhaps counter-intuitive mathematical point, but it is raised here because the choice of convention has an important practical implication. If the general UK convention (using internal dimensions) is used, the psi-values of all building junctions have to be determined and incorporated into the energy-performance calculations in order to model the building accurately. As UK Building Regulations are updated such that psi-values need to be incorporated into SAP calculations, clearly it would be impractical to require so many psi-value calculations to be undertaken for each building scheme. The alternative is to use accredited junction details that have known internal psi-values (ψ_i). By contrast, if the Passivhaus external psi-value convention is used, well-designed building junctions, an example of which is shown in Figure 8.4 on page 119, will effectively be thermal-bridge-free (if their ψ_e is 0.01W/mK or lower. As noted above, ψ_e can even be negative).

Calculating psi-values, whether internal or external, is complex and time-consuming. It is therefore better to use a convention that, where junctions have been appropriately designed, minimises or eliminates the need for these values to be calculated separately. This is why Passivhaus chooses to use external psi-values.

Assumed thermal bridge value at window edges in a Passivhaus

As we saw in Chapter 7, an installation psi-value ($\psi_{installation}$) has to be entered into the PHPP to reflect the additional losses between the window frame and the wall. (This is discussed further in Chapter 11.) If no thermal bridge calculations have been done, and provided the window installation detailing can be shown to have been adequately addressed, the Passivhaus Certifier will require a conservative $\psi_{installation}$ of 0.04W/mK to be entered into the PHPP. If the detail looks good, the Passivhaus Designer may be able to argue for a lower assumed value. However, it is better to design assuming a $\psi_{installation}$ of 0.04W/mK and to aim for an overall annual specific heat demand of 12kWh/m².a or 13kWh/m².a. This avoids the risk of having to do additional thermal bridge calculations to try to scrape through at just below the 15kWh/m².a limit for Passivhaus Certification.

Strategies for dealing with thermal bridges

Clearly, then, it makes sense to design out or minimise thermal bridges. As explained above, achieving this requires careful design or 'architectural detailing' of building junctions. Currently, however, there are very few, if any, off-the-shelf architectural details that designers can use to create thermal-bridge-free junctions. The Austrian publication *Passivhaus-Bauteilkatalog* [*Details for Passive Houses*], also known as the IBO book (see Resources), which is written in both English and German, provides helpful guidance and many examples. These details show the psi-values associated with them, which can be useful when assessing whether your own details are broadly acceptable. However, some of them are for types of construction that are not currently common in the UK and other English-speaking countries. And, even if they were, the book does not provide a 'copy-and-paste' solution for designers. In reality, there are many variables specific to any given project, which would often make straight copying of solutions impracticable. There are also useful example details contained in the **AECB Gold Standard** design guidance (see Resources).

Thermal bridging in retrofits

In most retrofits, some thermal bridging will be unavoidable. Where the building is to be externally insulated, there will be thermal bridging at the floor–wall junction, because there will always be a gap between the under-floor insulation and the external wall insulation. There is nothing that can be done about this, short of adopting extreme measures such as propping up the walls and replacing sections around the floor perimeter with an insulation material that has suitably high compressive strength. In most projects, this would make no sense financially.

Thermal bridging is also likely to be a problem at the roof–wall junction. In both cases, a bespoke architectural detail will be needed to minimise it. Figure 8.4 opposite, from the Totnes Passivhaus retrofit project, which is illustrated in the on-site photograph on page 120, show how the potential roof–wall junction thermal bridge was eliminated. In this case, the roof was new, the building having been extended upwards; however, the wall construction was a continuation of the existing twin-leaf cavity wall (using concrete block work).

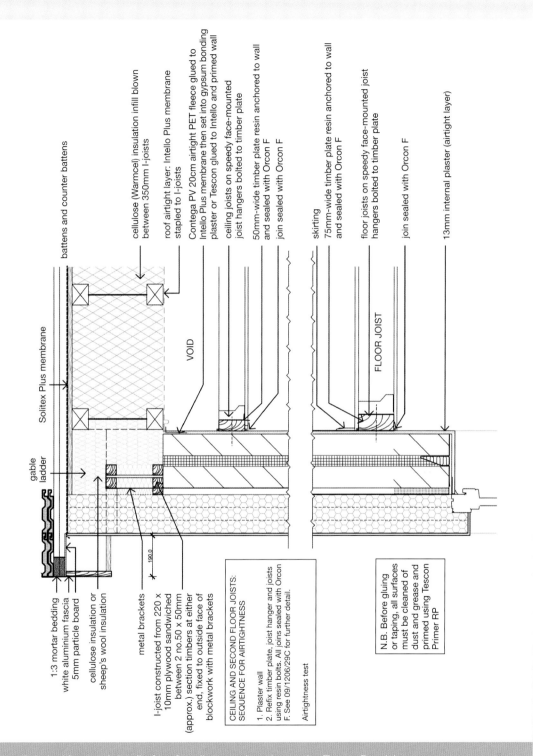

Figure 8.4 Architectural detail of roof–wall corner junction at the Totnes Passivhaus.

View of roof–wall corner junction from above, shown in the architectural detail on the previous page.

In a retrofit, the conflict between structural demands and the need to minimise thermal bridging is even greater than in a new build, because the building will not have been designed with thermal bridge avoidance in mind. A retrofit not only demands more of the architect and the Passivhaus Designer but also requires a pragmatic approach. This is reflected in the PHI's EnerPHit standard (see Chapter 1).

Calculating thermal bridges

If your design necessitates the calculation of a thermal bridge, there are a few eye-wateringly expensive software packages on the market designed to model even the most complex thermal bridge junctions in 3D. However, these are beyond the scope of this book and are intended for use by specialist mechanical and electrical (M&E) consultants; in any case, such software is overkill for the types of thermal bridging problems faced in most domestic- and community-scale building projects.

For architects and Passivhaus Designers wishing to do their own thermal bridge calculations (bravely or foolishly, depending on your view!), there is currently one quite old but free piece of software called THERM,[2] which can be used to calculate figures from which psi-values can then be calculated. It is used in the USA for calculating the thermal performance of windows, but it can be applied to other building elements. Although the software is free, it is wise to spend money on a one-day course[3] to learn how to use the output from THERM to calculate an accurate psi-value in accordance with the relevant EN standards. Although the software was designed very well in terms of its efficient use of computing power, it is not at all intuitive, so the course will save time spent on a slow and frustrating learning curve.

Once a psi-value has been calculated, this can be entered into the PHPP along with the length of the thermal bridge in metres; this allows the additional heat loss associated with the thermal bridge to be incorporated into the PHPP's energy model.

RECAP

Thermal bridges are caused when a gap in insulation allows heat to 'short-circuit' or bypass it. This occurs when a material with relatively high conductivity interrupts or penetrates the insulation layer. Most buildings have multiple thermal bridges because of the need to build for sufficient structural strength (constructional thermal bridges) or because of corners (geometry thermal bridges). However, with care, in a new build, it is possible to build a structurally sound building and also to eliminate thermal bridging. In retrofits, while elimination of all thermal bridging will be impractical, the same knowledge can be used to minimise it.

Most thermal bridges (whether caused by the building's construction or its geometry) are linear and are thus quantified by a psi-value. Some thermal bridges are measured as points (quantified by a chi-value): these are almost always caused by construction rather than by geometry.

Once the concept of thermal bridging is understood, the habit of addressing construction details to avoid it can quickly become part of normal working practice. With experience, you will develop knowledge of acceptable compromises where elimination is impossible. In the end, designing without thermal bridges is simply good common sense and is good practice in terms of avoiding moisture issues (by eliminating cold surfaces) as well as making the building thermally efficient. It also makes it easier to model a building's energy use in the PHPP accurately, as no separate calculations are needed.

CHAPTER NINE
Airtightness and sequencing

Air leakage and Passivhaus, internal air quality (IAQ), airtightness standards, wind-tightness, breathable materials, airtight materials, air leakage at the design stage and construction stage, sequencing, on-site communication and training, airtightness testing, typical airtight construction details

Perhaps the most debated aspect of the Passivhaus standard is its commitment to extremely low air leakage rates. This causes real concern for some people, particularly regarding issues of air supply. This chapter looks at the requirement and explains why this is adopted and how air quality and air supply are addressed. It reviews the important subject of the airtightness standards being proposed for the UK zero carbon standard (see also Chapter 5) and how these relate to Passivhaus. It looks at the practical challenges of achieving low leakage rates, and at typical Passivhaus construction details and how these can affect normal work sequencing.

Air leakage and Passivhaus

The Passivhaus standard stipulates exceptionally low air leakage levels, commonly referred to as airtightness. This is defined as the unplanned movement of air through the building's thermal envelope (walls, floor and roof, including any party walls). To meet the standard of below 0.6ach (air changes per hour) at 50Pa (pascals – the unit of force per unit area) above and below ambient atmospheric pressure, you are aiming to achieve leakage paths equivalent to less than the area of a 5p coin over every 5m² of envelope area. The main benefits of reducing air leakage are:

- saving of heat energy lost along with the escaping air
- improved performance of the insulation
- better comfort levels
- durability of the building fabric (minimising vapour moisture entering the fabric).

In a typical standard-build house, 15 per cent of heat loss can be attributed to escaping air, but it is in the improved performance of the insulation that the greatest energy benefit lies. Studies have indicated that even with the smallest of air leakages into the insulation layer (via gaps in the airtight layer of only 1mm wide), the insulation's thermal performance is significantly reduced (in one experiment, for example, performance altered from a U-value of 0.3W/m²K to a U-value of 1.44W/m²K – a factor of nearly five).[1] This is something that has not been properly appreciated by the building community and lies at the heart of the success of Passivhaus in delivering low energy performance, not just design intent. The failure of our low-energy buildings to perform as designed is widely recognised in the UK, and this failure is linked to a fundamental misunderstanding – simply installing insulation is not sufficient to guarantee performance. It must not be forgotten, however, that the third reason for reducing air leakage is simply comfort, i.e. the elimination of draughts and cold surfaces.

Achieving low air leakage levels has been demonstrated to be possible with all types of construction, from lightweight (timber) to massive (masonry). Success depends on ensuring continuity of the airtightness layer and on careful workmanship. Some would argue that wet constructions (masonry/concrete) are easier to make leak-free, as the airtightness solutions will be more tolerant of poorer workmanship – a concrete box perhaps being the easiest solution. However, choice of construction method and materials is influenced by a wide range of factors, and the Passivhaus standard deliberately does not favour one construction type over another.

Airtightness and indoor air quality (IAQ)

There is a degree of natural scepticism about achieving good indoor air quality (IAQ) while also achieving 'airtight construction'. A general anxiety remains that occupants might be left gasping for air! It would therefore be more helpful to describe such construction as 'leak-free', since most of us can understand the discomfort of a draughty room, especially on a cold day. In a cold and draughty house, we are inclined to avoid sitting next to windows and instead will gather around an open fire to enjoy the radiant heat. The high air change rates experienced in our typical buildings (an average air permeability of 7m³/hr/m² at 50Pa for UK new housing) might exceed what is necessary for our fresh air needs, but the disadvantages of such high rates are accepted as normal and tend not to be consciously acknowledged or considered unavoidable. For many, it is likely that only the experience of a true Passivhaus building will allay their nervousness about what is an unfamiliar approach. In a Passivhaus, fresh air is introduced by a simple mechanical method (see Chapter 12) and applied to the whole building in an integrated and controlled manner, ensuring that good air quality can be maintained throughout while avoiding oversupply (draughty) or undersupply (stuffy or smelly conditions). In contrast, using uncontrolled infiltration of air, such as trickle vents in windows, means that ventilation levels in the house rise and fall as wind speeds and temperature differences fluctuate, and that some spaces may be oversupplied while others remain undersupplied.

A key concern when deciding on a low-energy ventilation strategy is to ensure that potential pollutants in the indoor air are kept to a minimum. There have been very few academic studies on air quality in the home, but there is a growing body of concern about the possible link between the increased use of chemicals (glues/adhesives, sealants, varnishes, paints, carpets, fabric treatments, insulations, etc.) and an increase in the incidence of respiratory and skin disorders. Our older houses had no planned ventilation strategies; there were no window trickle vents or mechanical bathroom extraction systems – the air was purged and replenished simply through the general leakiness of the fabric. This was good for the level of most pollutants, but bad for energy consumption – assuming you want to live at a minimum of 20°C – and not great for comfort levels either. More energy-efficient (less leaky) buildings are necessarily reliant on planned ventilation strategies to maintain IAQ. In Chapter 12 (page 192) we refer to a study that demonstrated that much lower carbon dioxide levels were recorded in a Passivhaus apartment block with mechanical ventilation than in an older apartment block that relied on natural ventilation. (Carbon dioxide is an accepted marker for the wider mix of potential indoor air pollutants.)

Volatile organic compounds (VOCs)

Volatile organic compounds (VOCs) are chemicals that become a gas (evaporate) when at room temperature; a process referred to as 'off-gassing'. Commonly used building materials off-gas and thereby introduce contaminants into the indoor air. Extensive measurements have shown that the air in a typical home contains a large range of these VOCs. Most off-gassing happens at the early stages of exposure of the materials to room temperature, so leaving windows and doors open for a period before occupancy will help to significantly reduce such contaminants. Perhaps two to four weeks' 100-per-cent fresh-air flush should be incorporated into build programmes, especially if using 'off-gassing' products. However, products can continue to off-gas for years after installation, albeit at a lower level. We would therefore advise sourcing low-emitting or non-emitting products for an ultra-low-energy build – look for words like 'solvent-free', 'water-based', 'low-VOC' or 'non-toxic'. Products buried within walls, i.e. not exposed to air, are less of a concern, but surface finishes should be carefully considered. The fitting and using of any such materials can be very unpleasant for installers, and it is clear from having quizzed workers on-site that the matter should be a serious consideration. Paint products are already being improved as a result of EU legislation, so the concerns highlighted here are now being officially acknowledged. These issues are of, course, part of what drives people to adopt a philosophy of using only 'natural' building materials, as discussed in Chapter 5.

Airtightness standards

The measurement of air leakage is done using pressurisation tests during the construction period. This involves both pressurising and depressurising the interior of the building using large fans, normally attached to a door or window opening (more details on page 137).

While the test and kit used is common to Europe, the USA and the UK, the units of measurement vary and can cause confusion when making comparisons across projects or standards. The Passivhaus airtightness requirement of 0.6ach is measured at 50Pa, so the airtightness test applies a pressure of 50Pa above and below ambient atmospheric pressure against the interior of the walls of the house. This pressure is equivalent to a depth of 5mm of water applied over the surface of the building, and the air changes per hour at this pressure is often referred to as the n_{50} value. This refers to the flow rate of air entering and exiting the building (in m³/hr – cubic metres per hour) divided by the ventilation volume, V_{n50} (the total internal air volume, in m³).

This n_{50} value is different from the figure currently used in the Standard Assessment Procedure (SAP) calculations in the UK. SAP calculations are required to meet the UK Building Regulations (Part L1A for new dwellings) in England and Wales. The SAP figure is usually referred to as the q_{50} value and is measured in m³/hr/m², also at 50Pa above and below ambient atmospheric pressure. This refers to

Site blower test. A large fan, attached to the door frame, is connected to specialist software used to calculate air permeability and air change rate.

the cubic metres volume of air entering and exiting per hour per square metre of thermal envelope area (walls, roof and floor and including party walls if semi-detached or terrace), and is what is described as the air permeability of the building. The area of the internal surfaces excludes the surfaces within window and door reveals, and, while for this reason it is generally agreed not to reflect accurately the real air infiltration rates, it does allow useful comparison between projects.

The Air Tightness Testing and Measurement Association (ATTMA) is the main professional body whose members are qualified to carry out airtightness testing in the UK. They will measure the q_{50} value. To change the q_{50} value to an n_{50} value for Passivhaus, you need to multiply q_{50} by the thermal envelope area (measured as noted above) and then divide this by the internal air volume of the house. It is often quoted that $1m^3/hr/m^2$ at 50Pa equates to approximately 1ach, but this rule of thumb strictly applies only to a particular volume, equivalent to 6m x 6m x 6m, and therefore should not be universally applied. For other building volumes, discrepancies between measurements of air permeability (q_{50}) and ach (n_{50}) can be significant. Remember that the q_{50} value is a function of the enclosing surface area, while n_{50} is a function of volume. Often the surface area and volume are similar numerically, but not always!

Note that to obtain a value for air changes per hour under normal conditions (typically 4Pa air pressure difference), i.e. not at the test pressure of 50Pa, a useful guide is to divide the n_{50} value by 20. If the site is particularly sheltered, it is often advised that it would be more accurate to divide the n_{50} value by 30, and if the site is very exposed, by 10.

Current typical air permeability for new houses in the UK is $7m^3/hr/m^2$, while current (2010) UK Building Regulations (England and Wales) are set at $10m^3/hr/m^2$, so we can see that the step change required for the Passivhaus standard is quite significant. With an air permeability of $7m^3/hr/m^2$ we are throwing away significant heated volumes of air!

Airtightness standards compared

It is interesting to compare standards for airtightness internationally as well as within the UK – see Table 9.1 opposite, which shows a selection of statutory and voluntary standards. The highest statutory standard is in Sweden, and it is in these colder European countries that the value of airtightness has been recognised for some time, mainly for improved comfort levels. Countries with milder climates have, until recently, shown little, if any, interest in achieving airtight construction. The new motivation for more stringent statutory standards is less to do with comfort and more to do with energy use, but the comfort benefits are already well appreciated.

The question of why Passivhaus sets the airtightness target at a maximum of 0.6ach is of interest. Both the current UK zero carbon standard and the UK Energy Saving Trust best practice advise an air permeability (q_{50}) of $3m^3/hr/m^2$ – an air leakage limit where non-mechanical ventilation solutions can still be applied. Below this figure it is considered necessary to introduce planned ventilation solutions, typically involving a form of whole-house ventilation with heat recovery. By setting future regulations at $3m^3/hr/m^2$, building solutions using only natural ventilation techniques remain viable (normally some form of 'passive stack ventilation', driven by natural vertical pressure differences). These natural techniques will generally use standard window trickle vents to draw in fresh outside air. This approach carries with it an energy penalty, as a result of both higher leakage rates and the absence of heat recovery technology. We

Table 9.1 **International airtightness standards**

	Statutory/ voluntary	Standard	Test	Equivalent air changes per hour for an average-sized house
France	statutory	0.8-2.5m³/hr/m² @ 4Pa	q_{50}	11ach
UK Building Regulations 2010 (England and Wales)	statutory	Max. 10m³/hr/m² @ 50Pa	q_{50}	10ach
USA	statutory	Max. 1.6m³/hr/m² @ 4Pa	q_{50}	8.5ach
Switzerland	statutory	Max. 0.75m³/hr/m² @ 4Pa	q_{50}	3.3ach
UK zero carbon	statutory from 2016	Max. 3m³/hr/m² @ 50Pa	q_{50}	3ach
Sweden	statutory	Max. 0.8 l/s/m² [litres/second/m²] @ 50Pa		2.88ach
Canada Super-E	voluntary	Max. 1.5ach @ 50Pa	n_{50}	1.5ach
Passivhaus	voluntary	Max. 0.6ach @ 50Pa	n_{50}	0.6ach

Adapted from Limb (2001)[2]

look in more detail at heat recovery ventilation in Chapter 12, but at this point it is sufficient to note that an efficient heat recovery system, properly designed and installed, can recover more than 90 per cent of the heat energy from the exhaust air.

It is notable that the UK's zero carbon standard for new homes, currently still under review, includes a space heating target of 39-46 kWh/m².a compared with the Passivhaus standard of 15kWh/m².a. Given the target leakage rates of 0.6ach for Passivhaus and (currently) 3m³/hr/m² for UK zero carbon, the direct link between target leakage rates and energy in use is clear. It is also worth remembering that the method for measuring the treated floor area (TFA) is not the same in the PHPP and SAP, which is the current basis for the zero carbon targets, so the square-metre floor areas used in the space-heating targets are not directly comparable. SAP has more favourable rules for measuring a building's floor area (see Chapter 7, page 94), which means that the zero carbon 39-46kWh/m².a is actually equivalent to over 50kWh/m².a in Passivhaus terms.

The adoption of this lower airtightness standard in the UK in part reflects a cultural unease, as noted at the beginning of this chapter, with the concept of introducing fresh air into our homes by mechanical means, i.e. fans. However, most of us are experienced in managing what is known as 'mixed-mode' ventilation without perhaps realising it – in our cars. We open the car windows on bright, sunny days and enjoy the fresh outside air, while on other days (in heavy traffic, or in colder, wetter weather) we close the windows and rely on the clean, filtered, mechanically introduced air. In reality, mechanical ventilation systems are simple, relying on few working parts and offering many benefits in terms of internal air quality.

Whole-house mechanical ventilation is integral to the Passivhaus approach (see Chapter 12), and the airtightness standard needs to be below 1.5ach at 50Pa for this to be an efficient solution, i.e. generally recovering more energy than the Passivhaus-certified MVHR unit uses. The 0.6ach at 50Pa target allows for some loss of airtightness over the lifetime of the building without any undue effect on overall performance (e.g. on MVHR performance). The target also ensures there are no draughts with a velocity of more than 0.1m/s (metres per second), which is the speed above which we would physically begin to notice them. The effect of this non-draughty environment is argued to result in occupants setting thermostats to a lower air temperature (by as much as 2 degrees), since they feel warmer. This is a further example of Passivhaus functioning as a comfort standard as well as an energy standard.

Wind-tightness and thermal bypass

Passivhaus differentiates between airtightness and wind-tightness. Normally there is a designated wind-tight layer on the external side of the construction, coupled with the airtightness layer on the inside of the construction. The wind-tight layer acts like a protective wind-cheater, which improves the thermal performance of your woolly jumper by keeping air out. Reducing 'wind-washing', i.e. the penetration of air into and around the insulating fabric from the outside, improves the performance of the insulation. A good example is external insulation applied to a masonry substrate. Traditionally the fixing might have been carried out using 'dabs' of glue, but for Passivhaus a continuous layer of glue is required. The movement of air in and/or through an insulation layer is referred to as 'thermal bypass' (see Chapter 8, page 112). The air may travel through the fabric and return to the outside without affecting airtightness as such; the detrimental effect is in the lowered performance of the insulation itself. Insulation is not fully effective unless it is installed without gaps, voids or compression, and is aligned with a continuous air barrier (any durable material that restricts airflow).

Breathable versus airtight

A common misconception arises from the difference between the 'breathability' of construction materials, i.e. their ability to allow water vapour through them, and the airtightness of the construction. It is equally possible to build to the Passivhaus airtightness standard using either materials that are **vapour permeable** ('**vapour-open**') or those that function as **vapour barriers** ('vapour-closed' materials). This is because a material can be impervious to mass air flow – the movement of the body of air as a whole – but can still allow gases (vapour) to move by **diffusion** ('percolate') through it. This diffusion is driven by the difference in the pressure of vapour – the **vapour pressure differential or gradient** – on each side of the membrane. In an occupied building, there will be a higher concentration (higher vapour pressure) of water vapour on the inside, so the vapour pressure gradient will encourage water vapour movement from inside to outside. 'Vapour permeable' or 'vapour-open', then, in a construction context, is usually taken to mean 'permeable to water vapour'.

Materials also vary considerably in how vapour-open' they are: we usually refer to this characteristic in terms of vapour permeability levels. Some helpful illustrations are given in the next chapter. The different units used to measure vapour permeability can cause confusion, so ensure that like-for-like measurements are being used if you are making comparisons between products. A good example of an airtight but **'breathable'** or 'breather' material is Gore-Tex®, commonly used for walking and climbing clothes – this keeps the wind (air) out while continuing to allow moisture vapour from your body to escape to the outside.

An example of a vapour-closed or impermeable material is a rubber membrane (perhaps functioning as the watertight layer under a living roof). One example of a vapour-open material often used in buildings is woodfibre board, which might be included towards the outside of a timber-frame wall, where transfer of moisture to the outside is important. When using natural materials, such as timber, careful management of moisture levels within the construction is essential to ensure that materials can dry out if necessary. Water could be present due to a temporary accidental ingress (it does happen!), or materials could have become wet during the construction period and subsequently need to dry out. Traditional constructions have generally allowed moisture to be present, but also ensured that drying out occurred. Cyclical wetting and drying need not have detrimental impact, for example, in the case of rubble/stone solid walls. It is the trapping of moisture within constructions that can have the most catastrophic consequences. It is good practice to encourage moisture to move from the inside to the outside of the construction by using successively more vapour-permeable materials. As a useful rule of thumb, the vapour permeability of the material on the outside should be five times more than that of the material on the inside.

Intelligent breather membranes

The latest innovations in permeable materials are **'intelligent' membranes**. These breathable sheet materials have varying vapour permeability characteristics, depending on the relative humidity (RH) levels on each side of the membrane. This means that the membrane will be more vapour-open during the summer months, for example, when the average humidity level at the membrane is high – so if there is moisture trapped within the assembly, this can allow some drying out to the inside of the construction. Alternatively, if the average humidity level at the membrane is low (e.g. during winter months), the vapour permeability of the membrane decreases, protecting the fabric from moisture/vapour penetration into the construction and reducing the risk of **interstitial condensation** (condensation that occurs within a building assembly – see Chapter 10, page 147).

Intelligent breather membranes can be used as the airtightness layer for both wall and roof constructions. If used as the airtight layer within a timber assembly, they should be stapled frequently (say, every 100mm). The overlapped joints between sheets are then taped with an appropriate airtight tape. If counter-battening over the membrane (say, to form a service void), fix the counter battens over the top of any taped joints for a very robust solution.

Airtight materials

As noted, the airtightness layer is normally located towards the inside of the construction; this is often within a service void, for protection. The airtightness layer stops unwanted air leakage from the interior to the exterior, as well as helping to limit airborne moisture from the interior migrating into the structure, with the potential to cause deterioration to materials and decreased thermal performance. Moisture within the fabric of a building is the single most common cause of material degradation and building failure. Most insulation products are poor at stopping airflow. An effective air barrier is contiguous (continuous) across the entire building envelope, with all holes and cracks fully sealed, and should be in full contact with the insulation (referred to as 'fully aligned').

The most commonly used airtight materials are gypsum plaster or oriented strand board (OSB) of sufficient quality and thickness. Metals such as steel beams are (obviously) airtight, as is most concrete, as long as it is of a suitable quality. Ground-floor concrete slabs are therefore commonly used as part of the airtight layer, but this will need to be monitored for quality. Various membranes or sheet materials are also airtight, such as EPDM (ethylene propylene diene monomer) rubber-based sheets, polythene, polyethylene and 'intelligent' membranes. Table 9.2 below lists the relative airtightness of different materials.

Addressing the junctions between construction elements – such as where windows meet walls, and walls meet roofs, etc. – is key to achieving airtightness. These places are often where the airtight layer changes from one material to another, and most commonly such transitions are handled using a variety of airtightness tapes. The lifespan of these tapes is sometimes a cause for concern, and is yet to be 100-per-cent demonstrated, as buildings using such tapes are all relatively recent. The first Passivhaus, built in 1991, is still performing as intended, demonstrating a tape life of at least 20 years. There are manufacturers of such tapes who guarantee a minimum life of 60 years – and airtightness products have obtained Agrément Certificates (see box opposite), part of which would involve accelerated ageing tests. There is at least one Irish Agrément Certificate that covers an airtightness system including tapes and which concludes that the airtight system will have a life comparable with the other elements of the construction.

It is best practice to protect these tapes with a covering layer to avoid future damage. Often tapes are part-buried within the plaster layer,

Table 9.2 **Materials that are suitable/unsuitable for airtightness layer**

Airtight materials	Non-airtight materials
Airtightness tapes and grommets	Cement
Concrete (quality-dependent)	Chipboard
Glass	General-purpose sealants and foams
Good-quality OSB (18mm or thicker)*	Insulation
Gypsum plaster	Masonry (bricks or blocks)
Intelligent breather membrane (vapour-open)	Plasterboard
Lime plaster (to suitable specification)	Tiling and grout
Steel	Unprocessed wood
Vapour barrier membranes (vapour-closed)	Woodfibre boards

* But see box opposite

> ### Agrément Certificate
>
> This is an approval system awarded to a product after it has successfully passed a comprehensive assessment involving laboratory testing and which includes continued monitoring of the product after the certificate is granted. Various countries, including Britain, Ireland, France and Germany, run these product-approval systems.

which will further ensure longevity. Similar glues to those used in airtightness tapes are employed in other industries in far more aggressive and demanding environments, so there is good reason to trust that the tapes will perform in the long term. Should the airtightness deteriorate a little over time, a Passivhaus should still continue to operate as designed, and certainly better than required by the current zero carbon standard, although there would be some small energy penalty. The mechanical ventilation would continue to operate effectively at up to 1.5ach.

Airtight tapes are not always easy to stick into place, and if this is found to be a problem it will normally be related to dust (use a primer first) or moisture levels. If the surface of a membrane is damp, for example, adhesion will be problematic and the surface will need to be dried first (a hairdryer can be useful!). If possible, avoid using the tapes while the atmosphere is damp.

Neither brick nor block walls are airtight, and typical mortar joints will often prove to be very leaky. The tendency for long-term deterioration of mortar joints (eventually requiring repointing) means that careful consideration is required if intending to retrofit older masonry buildings to any level of airtightness. For wind-tightness, repointing of the external mortar joints is likely to be necessary if insulation is to be applied internally. This is to avoid both thermal bypass and rain penetration (especially if on an exposed site). Traditionally, the surface area of wall

> ### OSB airtightness
>
> Oriented strand board (OSB) is made airtight by taping along the joints between adjacent boards using airtightness tape. The normal recommendation is that OSB should be a minimum of 18mm thick. However, it has been discovered that some boards at this thickness are not airtight – this probably reflects some variation in board composition as supplied by different manufacturers. A simple blower test prior to use (use a balloon taped against the board surface and blow air against the opposing side) will confirm whether the board is up to the task. In some projects the boards have had to be lined with an airtight membrane after the problem was identified post-installation. A better recommendation might be to use OSB/3 grade boards and maybe at 22mm thickness. For those who are interested, there is a useful study on the air permeability of eight different brands of OSB.[3] The study reveals a wide range of permeability, even within same brand boards, and suggests a shift to using a recycled-wood-based panel instead of the OSB, with an additional airtight top layer (e.g. a membrane or foil).

located within the internal floor zone (above the ceiling and below the floorboards) was never plastered, so this will be another particularly 'leaky' area. With Passivhaus, the plastering of these previously hidden areas is a significant change from traditional building practice.

It is important to ensure that inappropriate materials are not used on-site instead of more expensive and specific airtight tapes and glues; in particular, polyurethane foam or silicone sealants are not acceptable and will not be durable over time.

Avoiding air leakage at the design stage

Identifying the airtightness layer early on in the design stage is critical, as the construction methods and materials to be used will affect which airtight materials are appropriate. Key drawings highlighting the airtight layer (often blue or red are used as a colour code) should be prepared – normally by the architect – for use by the construction team. An example is shown in Figure 9.1 below.

The airtightness layer should be continuous and run around all the building elements (floor to wall; wall to roof, etc.). Each transition point, usually at construction junctions, will need to be detailed separately (as in Figure 9.1) so that the site team is clear about how the airtight layer transitions from one material to another, or wraps around elements such as intermediary floors. This process will soon highlight the fact that there is a need for particular sequencing of the work in order to achieve this continuity. It is therefore helpful if any unconventional sequencing can be highlighted on the construction drawing, as in Figure 9.2, page 135. And, although a wide range of tapes is available, each for a specific situation, it is often simpler to limit the number of tape varieties specified to avoid the potential of misapplication on-site.

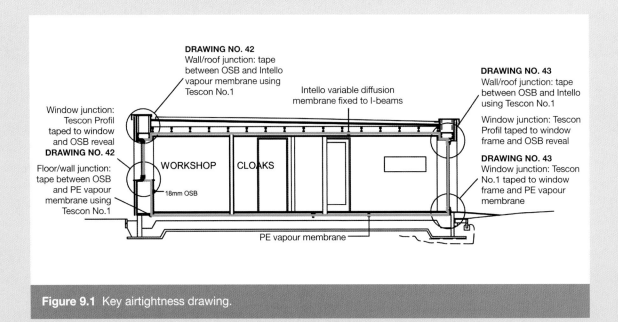

Figure 9.1 Key airtightness drawing.

Be practical about the location of sockets, using internal rather than external wall positions if there is a choice (in a retrofit, avoid the first 1.2m from external walls). The need for service penetrations through the external envelope should be kept to a minimum. Often a 50mm service void is created on the inside of the airtightness layer. Simplifying the junction details and avoiding too many different transitions between airtight materials will also help.

During construction, the need for power cables to run a variety of tools creates a risk of additional, unplanned holes through the fabric. It is also true that future services might require penetrations through the fabric after the building is complete. It is therefore suggested that at least one spare entry point is designed in, and noted on the drawings. A single grommet with multiple entry points could be used.

If formally tendering to contractors who have no experience of building to high airtightness standards, it would be wise to meet face-to-face with any of those being considered and to specifically discuss the airtightness issues with them. Decide how realistic they seem to be about the challenge being set them, and how well informed they are. The airtightness standard is a challenge for the construction team, and no amount of design detailing will solve poor attention to detail on-site – even if it is carefully hidden from the eye! Ensure that the expectations and targets are clearly documented within the tender, including the provision of the airtightness tests and the requirements for on-site training or induction sessions relating to airtightness. Passivhaus Certification also requires photographic records to be made during construction. Make sure that this is clear to the contractor and that how it is to be managed is also quite clear. Taking good photographs will provide useful information for future training, in any case. Ensuring the right choice of construction team is discussed in more detail in Chapter 6.

Avoiding air leakage at the construction stage

The process of achieving an airtight building will soon highlight the need for close cooperation between the different trades. The need for a well-managed and integrated workforce cannot be overemphasised; careful workmanship and tight supervision are the key requirements.

The commonplace tendency not to provide on-site workers with a broad-based briefing (the 'how and why') of the project in hand contributes to a build process that becomes a series of isolated interventions by workers aware only of their own tasks and not of the potential wider implications of their work or habits. This will cause havoc for those wanting to achieve an airtight construction, and part of the remedy is to ensure that proper induction sessions are held.

At the construction phase (on-site work), we would advise the inclusion of an initial induction session for the whole workforce. This can be useful to explain the overall approach of Passivhaus for those unfamiliar with it, as well as to run through the essentials relating to airtight construction. It is critical to ensure that key drawings (e.g. airtightness diagrams – see Figure 9.1 opposite) are always available for inspection on-site and that there is a clear understanding of which materials constitute the airtightness layer on each element of the building's envelope (wall, floor, roof, etc.). The tendency for different trades to follow each other in sequence (electrics, plastering, decorating, etc.) and to subcontract their work allows for 'grey' areas of responsibility to arise, and for blame to be shifted from one party to another. The key, then, is to find a way to help each group understand the final goal and the reasons for it, as well as help them appreciate that each has a significant contribution to make. It is important that the consequences of not delivering a contiguous airtightness layer are understood, including serious material degradation through moisture ingress (see Chapter 10). Poor practices such as the 'build–damage–install–repair' cycle, which occur too commonly on UK building sites, will prove enormously problematic if trying to achieve such demanding airtight levels. Hidden areas (under stairs, in floor voids, under baths, etc.), which are commonly very poorly finished, must be finished as effectively as other areas, and this must be communicated at the outset.

It is advised that an on-site 'airtightness champion' is appointed (this need not be that person's only role!), although no single person can guarantee that all mistakes are picked up – work can be swiftly concealed in the natural process of building. Therefore it is important to encourage the entire team to have a responsible attitude towards airtightness. Meeting the Passivhaus standard of 0.6ach need not be onerous if a whole-team approach can be achieved. Making walls, floors and junctions airtight is not, in reality, overly complex or difficult to learn. An informed workforce can easily be encouraged to think in airtightness terms and may soon be suggesting improved techniques and solutions. Any problems or proposals need to be reported to the airtightness champion so that proper feedback and ongoing learning is in place.

There are additional costs associated with achieving a continuous airtightness layer, in particular budgeting for extra internal plastering, where it is being used as the airtight layer. In retrofit projects, it would be wise to include the complete removal of the existing plaster in the budget, as a contingency, unless it is certain that it is completely sound. Some of the proprietary products available (e.g. airtight back boxes for electric sockets and grommets for cables that penetrate the airtightness layer) are not always essential, and on-site solutions can be perfectly effective (e.g. setting the electric socket into wet plaster when being fitted). Care must be taken with flat-profiled cables when using grommets with round holes – the grommet may not then secure an airtight finish.

The most common areas of weakness are door thresholds and the junctions between windows and reveals, so it is worthwhile paying additional attention to the workmanship in these areas. In particular, windows are often installed by a window subcontractor, who may not be used to fitting the expensive, high-performance Passivhaus windows essential to such a project. The thermal performance of these windows is very much affected by how efficiently they are installed in the window opening (see Chapter 11). UK installers are very unlikely to be familiar yet with the accuracy and care required and we would advise caution, even if using approved installers. The sequencing of and responsibility for window installation needs careful thought (see 'Installation of triple-glazed windows' opposite).

Building in tolerance for future movement – especially at junctions between materials – is necessary, although some tapes have a certain amount of tolerance within their design (typically around 10mm). The airtightness layer must follow the general principles of good building practice, and this is especially true where timber structures abut masonry construction.

Changes made during the construction period are notoriously difficult to manage, but are not always avoidable. If changes are made, the lines of communication need to be in place so that the architect/designer and the airtightness champion are fully informed, and can ensure that any impact on the airtight layer is managed. At the end of the project, make sure there is a post-completion review; this is such a valuable learning opportunity but is too often squandered.

Sequencing during construction of a Passivhaus

In standard buildings, we are used to various sequencing requirements – trades follow each other as work progresses, and it is hoped that a sensible ordering of activities ensures that any damage to previous work is kept to a minimum and efficiency is maintained.

However, the particular requirements of a Passivhaus build sometimes lead to the need for non-standard sequencing. Generally this will arise from the requirement to create a continuous airtight layer around the building shell. If the build progresses without a particular sequencing step having taken place, retrospective solutions around junctions may become impossible, expensive or time-consuming.

Sequencing is a question of understanding the problem at hand; the solutions are often simple and straightforward to execute. The challenge is essentially one of clear communication with all parties involved, especially on-site, otherwise operatives will simply plough on and complete work in 'normal' sequencing order.

When the construction method is being decided on and construction details prepared, it is helpful for the architect/designer to consider (as early as possible) if there will be a need for any unusual sequencing. Once identified, these must be clearly highlighted to the contractor/builder. Adding breakout boxes on drawings (see Figure 9.2, right) and then numbering the activities in the correct order is one helpful approach. Even better is having an initial discussion with the builder. Making use of their on-site practical experience is beneficial.

Examples of unusual sequencing

It is useful to look briefly at two of the most common examples of unusual sequencing: firstly the installation of triple-glazed windows, and secondly a situation where a potential conflict arises between structural and airtightness requirements.

Installation of triple-glazed windows

Window reveals (the sides of the wall in which your window will sit) in masonry constructions will need a **parge** coat – a thin plaster layer – prior to the window installation (pictured overleaf), which is not common building practice. This is to avoid air leakage into exposed masonry. (Remember that windows form part of your airtight layer.) There will also need to be agreement on which activity is to belong to which trade, in particular who is best made responsible for the application of the airtightness tapes around the frame perimeter. Would this be best carried out by the window fitter (likely to be a subcontractor) or by the main contractor? It is normally at points where work overlaps between parties that responsibilities become confused and mistakes occur.

Sequencing notes for airtightness

1. Plaster reveal in preparation for window installation.
2. Fix metal frame cramp to the outside of the existing wall.
3. Install window.
4. Stick tape to the inside of the window frame as per manufacturer's instructions.
   Embed the tape into a coat of plaster to create continuous airtight layer between plaster and window.

Figure 9.2 Example of airtightness sequencing highlighted on construction details.

On-site communication, site structures and on-site training

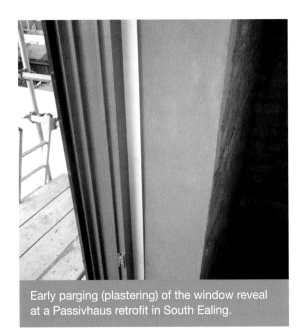

Early parging (plastering) of the window reveal at a Passivhaus retrofit in South Ealing.

As with the installation of the airtightness layer, unusual sequencing requires clear communication with all on-site workers, and it is important that this happens at an early stage. Different trades will be carrying out work in the same location at different times, and it can be quite demoralising to see your good work undone by another trade that did not appreciate either a sequencing or airtightness factor! Practical face-to-face communication is probably more effective with on-site workers (not relying solely on drawn information), especially as operatives are being asked to alter some of their usual working patterns, and at least initially are being slowed down – and naturally there will be some resistance to this.

Again, independent adoption of the principles of low-energy construction by the site workers is so important – a professional can never predict every consequence of a construction detail, especially any practical restrictions that might be involved. Intelligent problem-solving on-site is an invaluable skill, as long as it involves appropriate communication with the team before implementation. A full briefing of site workers is an important part of this empowering process.

As noted in Chapter 4, in the UK there are fewer and fewer trades employed directly by construction companies. Most tradespeople are self-employed and move from site to site as work becomes available, working either directly for a client or indirectly for a larger building company but on a self-employed basis. Electricians or plumbers, for example, will of course be very focused on their own element of the work and will have limited interest beyond this, especially

Airtightness and structure: floor-to-wall junctions

There can be particular conflicts between the airtightness layer and load-bearing elements. In typical constructions, beams and joists can penetrate the wall structure (bearing their load on to the walls), but once you are in low-energy design mode these penetrations will tend to interrupt the airtightness layer, and you need to ensure that the layer is maintained across the junction. In new builds, the most common approach would to pre-wrap the floor end with an airtight membrane. This can then be taped above and below the floor to the wall airtightness layer. In timber-frame houses, the floors can be pre-made off-site as a single element or cassette. The roof-to-wall junction can also be simplified in this way, using a similar cassette system.

The earlier you look at any such potential sequencing conflicts, the cleaner and simpler solutions are likely to be on-site.

if on a fixed price. The people they are working alongside will be continually changing as they move from job to job. In these circumstances, it is very difficult to introduce non-standard practices or to build up strong team working. This reality has to be faced honestly and then actively managed.

In order to address some of the communication issues already highlighted, we feel it is essential to include some on-site training; this can be a respectful way of engendering a team approach. It should give those working on the project a basic overview of ultra-low-energy principles so that the reasons for unusual sequencing or different detailing are clearly understood, along with the potential consequences of ignoring them. It is important that the implications of not delivering a contiguous air tightness layer are understood, including serious material degradation through moisture ingress and potential mould growth within the fabric, which can affect indoor air quality (see Chapter 10). Building to ultra-low-energy levels creates a different internal environment, which needs to be properly appreciated. People will be much more likely to change working patterns from a place of understanding than from being given disconnected (and what may seem superfluous) instructions.

Training needs to address both plumbing and electrical penetrations of the airtightness layer. When pricing for the job, both these trades need to understand what is required of them in this regard. The airtightness details are not in themselves difficult, and making penetrations airtight is relatively straightforward if carried out at the correct stage – again, sequencing needs to be understood from the start.

Governments have continued to introduce increasingly demanding legislation in many areas of our lives, but at present such low-energy building legislation remains at discussion/draft stages. Currently there is no access to free retraining either – for builders, architects, planners, building control officers, etc. (The limited training opportunities for construction workers are described in Chapter 4, pages 48-9.) In these circumstances, low-energy projects are likely to continue to be mainly generated by enthusiasts – but not all those working on such projects are likely to be as enthusiastic! Experience demonstrates that being offered intelligent explanations appeals to the majority of those who want to do a good job, who will enjoy learning (and being challenged) and who will then reap the satisfaction of a higher-quality build. If, however, people are frustrated with poor organisation and abortive work, that goodwill can soon dissipate. Good project organisation, clear communication and the provision of some training are the three most helpful tools in getting site workers fully on board.

Airtightness testing

An airtightness test requires some basic kit – a digital pressure gauge, a calibrated fan and an airtight door panel (see photo on page 125) – and the test needs to be carried out by an accredited person. In a typical house, a minimum of half a day is needed for a test to be carried out, including setting up, testing and checking the fabric. The airtightness test result will be affected by high winds, so wind speeds on the day need to be below 6m/s or 21.6km/h. The final test result is an average reading taken under pressurisation and depressurisation; the average must be below 0.6ach.

A minimum of two tests should be allowed for in the build process, but the ideal is to budget for three tests at domestic scale, especially if you are unfamiliar with achieving airtightness at this level. The first test would occur as soon as the external envelope is weather-tight and the doors and windows are installed. This can enable testing of a timber frame prior to any service penetrations, and allow early payment of this package of work, should it be a subcontracted element. The second test would occur once all service penetrations are complete but the airtight layer is still exposed. The final test is of course carried out on completion. The first two tests are opportunities to check that everything is performing as intended; mistakes can then be highlighted and any remedial action taken (obviously this is only true within certain limits). Once the team is practised at achieving airtight constructions, two tests should be sufficient.

Before the test, ensure that any intended leakage paths are sealed, e.g. ventilation units (seal off with tapes) and drainage traps (fill with water).

Gaps in the airtight layer are generally identified using smoke (e.g. smoke guns, pencils or tubes), although a thermal imaging camera can be useful in winter, if available. Under pressure the smoke provides a dramatic illustration of the leakiness potential, and will stream out of the building through any gaps! Of course you can always take the low-tech approach and use your hand (especially the back of the hand). As an example, on one test a single hole in the vapour barrier had not been taped after the roof void had been filled with wood-fibre insulation – the hole was where the pump nozzle had been inserted. This sort of error is exactly the type that can be quickly and effectively resolved at minimum cost. The pressure test is therefore an essential tool and must be carried out prior to any covering up of the airtightness layer.

For larger projects, the contractor might feel that a small airtightness quality control unit would be a cost-effective purchase. This is a smaller, lighter and uncalibrated version of the fan unit from a full testing kit. It is not able to give test results, but will be invaluable in testing areas as the work progresses, and is also a useful learning tool.

Typical airtight construction details

The images on the following pages show a range of common airtightness details, covering both retrofit projects and new builds. Once the principles of airtight construction are understood (by the whole team, and especially those on-site) it is relatively straightforward to amend and tweak these examples to suit the specifics of each individual project. Finding robust solutions is the key – membranes and tapes can get damaged, especially if they remain exposed on building sites for lengthy periods before being covered up.

Externally applied insulation, retrofit detail

Externally applied insulation on a masonry wall around a window frame. Joints should be staggered by 20cm so that air penetration is limited. Fill joints with supplied foam filler, and ensure the use of thermally broken anchor fixings.

Junction between timber floor and masonry wall, retrofit detail

1. Initial plastering (parging) of the floor void between the floor joists – the airtight layer is the gypsum plaster. This is non-standard practice; in a conventional build this void is left as bare block or brickwork.

2a. Joist ends are taped to initial plaster (parge) coat. Any dusty surfaces are prepared with a suitable primer prior to applying the airtightness tape. The final plaster layer is applied over the tape to complete.

2b. Alternative detail to 2a, with the integrity of the continuous plaster layer maintained by using a timber plate bolted through the plaster into the structural masonry wall using resin-bonded anchors. Floor joists can then be hung off the wall plate in a conventional manner using metal joist hangers. Plate edges can be sealed using airtightness sealant for the extra-cautious!

2c. Alternative detail to 2a or 2b, using a steel beam, which sits on the two party walls and allows all the timber floor joists to be trimmed short of the cold external masonry wall. The steel is also airtight, so could form part of the airtightness layer.

Airtight tapes and windows
A. Solution using only tape (the usual method described for windows or doors)

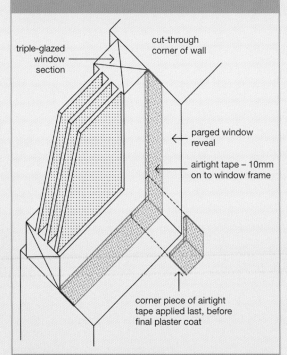

- triple-glazed window section
- cut-through corner of wall
- parged window reveal
- airtight tape – 10mm on to window frame
- corner piece of airtight tape applied last, before final plaster coat

1. Airtight tape is stuck to the face of the frame prior to fitting, using a tape with double-release papers – normally a width of 10mm can be released first to stick on to the frame. Mark the 10mm around the frame with a pencil – over-taping may cause you problems later when trying to cover up the tape with the finishing layers of your window reveal.

2. Tape is temporarily fixed back so you can fit the window easily (use masking tape or similar). Once fitted, you can release the second paper and adhere the tape to the airtight layer of your wall, e.g. pre-plastered (parged) window reveal or OSB.

3. Airtight tape is adhered to the window reveal around the perimeter of the window frame, using primer where dusty to ensure airtightness.

Chapter Nine • Airtightness and sequencing 141

4. Once the window is fitted and the other perimeter tape has been stuck down, you need to form and fit tape corner pieces. This photo shows a mock-up of a separate corner piece of tape on a spare piece of wood. Origami skills are useful here – these corners are the likely weak points with this approach. Again, the tape should extend no more than 10mm on to the frame so it can be covered at the final stage. We would suggest at the end applying some Orcon F® glue (or similar) in each corner, just to be quite sure of airtightness. The alternative is to use a membrane around the entire frame, as outlined in option B, right.

Airtight tapes and windows
B. Solution using membrane and tapes (this might be a more robust and flexible approach – especially for doors)

1. Initial stapling of the membrane to the window or door frame. The membrane is then taped to the sides of the frame with airtight tape.

2. Metal ties are screwed into place, ready for lifting up into the prepared wall opening. Once in position, the membrane can be taped around the opening to the airtight layer (whether masonry or timber) at a point to suit the build sequencing.

Intelligent vapour membrane on a timber roof

1. An intelligent vapour membrane used as the airtight layer on a timber roof, stapled frequently to the timber roof joists and then taped between adjacent sheets. Ideally fix the timber battens (to form the services void) directly over the taped junction for a very robust detail – as shown below.

2. Junction with masonry wall. Proprietary fleecy tape is stuck to the vapour membrane and the fleece embedded into the plaster on the wall. Leave some slack in the membrane in case of future movement.

Roof ridge detail

Sequencing of work is often key, and at the roof ridge the airtight layer must be continuous under the ridge beam. Allow for the early installation of a section of the airtight layer, and protect it from damage during construction by covering with a layer of ply. As a rule of thumb, fixing with nails is generally acceptable if fixing timber into timber.

Use of airtight grommets

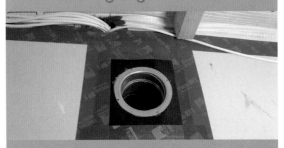

Proprietary airtight grommets: **Above:** for exit point of ventilation duct; grommet then taped around its edge to the gypsum plaster wall. **Below:** for cables through the airtight membrane.

RECAP

Passivhaus sets the most stringent airtightness standard in the world, 0.6ach at 50Pa, and this is applied alongside the use of mechanical ventilation with heat recovery (MVHR). The current zero carbon airtightness standard of $3m^3/hr/m^2$ leaves the door open to natural ventilation strategies, with significant energy penalties. Reducing air leakage is key to the Passivhaus approach for several important reasons:

- As air leaks out of your home, heat energy is also lost.
- Air moving through your insulation through gaps etc. reduces its effective performance.
- As air moves through your construction (walls, floors and roofs), moisture is also carried into the structure and this can have negative consequences, especially for material degradation (and mould may also affect indoor air quality [IAQ]).
- Low air permeability is essential for a mechanical ventilation system to operate efficiently.
- Reduction in draughts increases comfort levels.

The materials used to ensure airtightness can be breathable, i.e. vapour-open, and still achieve fantastic airtight results. Using such materials can help to contribute to a healthy IAQ.

Non-standard sequencing is likely to be required on-site to maintain the integrity of the airtight layer, which must be unbroken. Early identification of any sequencing issues is important. An airtightness test is needed prior to covering up the airtight layer and is invaluable in identifying small leaks; some remedial work is then possible, and at much lower cost than if left till later.

Achieving airtight construction on-site is a challenge both in terms of general building culture and also because it needs to be supported by appropriate training. If the principles of airtight construction and sequencing are communicated successfully to the site team, it can provide both an enlivening experience and a positive challenge that the contractor can grasp with both hands! Because airtightness is a physically tested quality, success can be a real source of pride for the construction team. While the architect contributes to setting up the airtightness strategy, it is the construction team that has the real task and challenge of meeting this significant part of an ultra-low-energy build.

CHAPTER TEN
Moisture

Liquid moisture and water vapour, relative humidity (RH) and indoor air quality (IAQ), capillarity, hygroscopicity, vapour permeability, moisture management in construction, breathability, example constructions (new build and retrofit)

Chapter Ten • Moisture

It is clearly not possible to make anyone an expert in moisture management in a single chapter of a book. This is partly because water/moisture is affected by so many different variables: climate, location, construction type and method, material, occupant behaviour, etc. But it is important to gain an overview of some of the main issues, especially on an ultra-low-energy build. Moisture levels both within the building fabric itself and transported by the internal air need to be carefully considered.

The Passivhaus approach is well considered when it comes to combining low energy performance with moisture management, and this is one example of how all the principles of Passivhaus together provide an integrated solution. Adoption of the standard needs to be as a whole – the temptation to partly adopt strategies, particularly in retrofit scenarios, carries significant risks. Highly insulating some areas while leaving others exposed will concentrate moisture towards localised areas, so a structure that could previously manage moisture levels may become overloaded. Highly insulating without ensuring low air-leakage levels increases the risk of moisture travelling into your construction and condensing (interstitial condensation – see page 147). Both these scenarios could lead to structural damage, a reduction in thermal performance and fungal growth (which can affect indoor air quality, or IAQ). However, you can also prolong the life of your construction materials by ensuring good moisture management – this is part of the quality approach built into the Passivhaus standard.

Moisture has been referred to as the worst of all potential pollutants in a building, and it is certainly identified as having the largest detrimental impact on the fabric of buildings, in terms of both structural and material degradation. The moisture (relative humidity, or RH) level in the air largely determines whether a variety of microbiological pollutants will be able to thrive, which has direct consequences for our health and well-being. The RH level also influences how comfortable or uncomfortable we feel (as it affects our ability to sweat and control our own body temperature) and will alter the way we experience a given temperature. We are more tolerant of higher temperatures, for example, if the RH level is low.

Unfortunately, the potential for these detrimental effects increases as we increase the level of insulation in our buildings. Traditional buildings were not heated to any great degree, and neither were walls, roofs or floors insulated. Moisture presence was managed by building with materials that could naturally dry out, and allowing air to pass through and around elements to aid this process. As we now expect warmer indoor temperatures, are increasing airtightness levels and using greater amounts of insulation, we are also creating different indoor environmental conditions. When designing ultra-low-energy housing, considering how moisture will be managed is therefore extremely important.

This chapter is intended to ensure that you approach the design of your building fully aware of the detrimental consequences of poor moisture management. In a new build, the risks are controllable – you can design them out. In a retrofit project, moisture management becomes more complex and your management strategy even more critical. The first part of this chapter describes the processes involved in moisture transport in buildings. The second part deals more directly with construction and discusses some examples of low-energy constructions.

Liquid moisture and water vapour

When we consider moisture, we usually think of it in its liquid form; the kind that falls as rain and hangs in the air as mist and fog. Liquid moisture has the potential to do the most damage to your building. But we also need to be aware of water in its gaseous form, water vapour. While invisible to the eye, water vapour is always around us and makes up approximately one to four per cent of the Earth's atmosphere. Moisture is transported by the air in the form of water vapour, and will readily return to its liquid state given the right conditions.

Water molecules are polar molecules (they behave a bit like mini magnets) as a result of the asymmetrical distribution of their constituent parts – hydrogen (positively charged) and oxygen (negatively charged). This means they tend to group together (in pools of water) and also adhere easily to some materials, such as smooth glass, and be repelled by others, such as wax. Some materials will soak up (and give off) moisture from the air much more readily than others, and in turn this will affect RH levels – the quantity of water vapour in a given volume of air – especially at low air change rates. With high air leakage rates, typical in most housing, the influence of materials on humidity is much less significant. The way a particular material interacts with moisture influences how moisture might or might not get transported through your building's thermal envelope. Poor moisture management within your construction will affect the durability of the materials, can encourage moulds to develop, and will often be detrimental to energy performance.

The difference between liquid water and water vapour is in their molecular density: water vapour is less dense than liquid water, while the molecular bonding present in liquid water also reduces significantly in its vapour form. Water vapour, like any other gas, will diffuse through a material if there is a higher concentration of the vapour on one side of the material than on the other. This is why we can have watertight sheet materials, often referred to as 'breathable' or 'breather' membranes, that are both airtight and liquid-moisture-tight but water-vapour-open, i.e. they act as a barrier to liquid water and to the body of air as a whole, but allow water vapour to pass through them (see Chapter 9, page 128).

Understanding this concept is key to appreciating how airtight constructions can also be 'breathable' constructions. When we refer to 'breathable' materials, we are normally considering how they interact with moisture. So with an airtight,

Monolithic films

Conventional airtightness membranes are normally microporous, transporting vapour 'passively' through their open pore structure. There are also membranes on the market that include an additional monolithic film (TEEE film). These are non-porous but allow water vapour to diffuse 'actively' through their molecular structure, by molecular exchange. Unlike microporous membranes, monolithic films have selective permeability – they allow only some, not all, gases to pass through. These films are usually hydrophobic, i.e. repelling liquid moisture. They might be used where a watertight membrane is needed towards the outside of an assembly and there is a risk of condensation forming on the inside of the membrane (say, in a timber roof construction) – this will help to prevent mould growth on the surface.[1]

low-energy construction, depending on the materials you use, water vapour may still enter and exit your construction.

Interstitial condensation

In winter, temperatures will generally drop across the construction depth (from inside temperature to outside temperature). It is therefore possible, with the right conditions, for any water vapour to return to its liquid state, i.e. to condense, on a surface within the construction. This is known as interstitial condensation. In this way, liquid moisture can accumulate within the fabric of a building over time. If this water is trapped, the detrimental consequences can be extreme. However, if the water can dry out (normally to the outside), any damage can be limited or can have no effect at all, depending on the materials involved.

If the moisture accumulation is the result of air leakage, you will tend to get localised areas of concentrated moisture. Moisture transported through vapour-open 'breathable' materials will distribute the moisture more evenly across the whole construction. Moisture penetration by air leakage is never planned and may cause localised failures. However, should moisture transported through vapour-open materials become trapped within the construction, failure could well occur across the entire construction element. One problem with moisture condensing within constructions is that the effects can go unobserved for long periods, perhaps coming to light only when the consequences have become severe.

Moisture in the air

If you do not achieve appropriate airtightness in a super-insulated building, then the internal air will transport moisture produced by occupants into your building assemblies, wherever there are gaps or cracks to allow it to leak in. Designing an airtight contiguous (continuous) layer is therefore critical in avoiding moisture transfer into your wall, roof or floor. Airtightness, then, is not only a strategy for lower energy use but is also an important part of the way in which a low-energy building manages moisture in a healthy way. Once your structure has been highly insulated, it makes no sense to allow warm, moisture-laden air to penetrate into it.

In terms of the external air, your wind-tight layer (see Chapter 9, page 128) will act in a similar way to protect your assembly from ingress of moisture-laden air. In cool-temperate climates, such as in the UK, the vapour pressure differential, or gradient, for most of the year will be from inside to outside, so the internal airtight layer is more critical. The vapour permeability of materials should increase towards the outside of the assembly to further encourage moisture transport to the outside (see 'Breathable assemblies', page 156). In the summer, when the outside air is warmer and moister, the reverse gradient is much weaker because of the smaller average temperature difference between inside and out, as well as the more rapid drying effect from wind and sun.

Absolute humidity and relative humidity (RH)

It is best if we define the term 'relative humidity' before proceeding further, since RH levels are generally what is referred to when considering ideal indoor air conditions. 'Absolute humidity' is an easier term to comprehend: it is simply the mass of water vapour divided by the mass of dry air in a given volume of air. Humidity is measured at different temperatures because temperature affects the ability of air to contain or carry water vapour – hot air can transport more water vapour than cold air (see Figure 10.1 overleaf). This is why cold air can sometimes be described as 'dry'. 'Maximum absolute humidity' means that the air is transporting its maximum

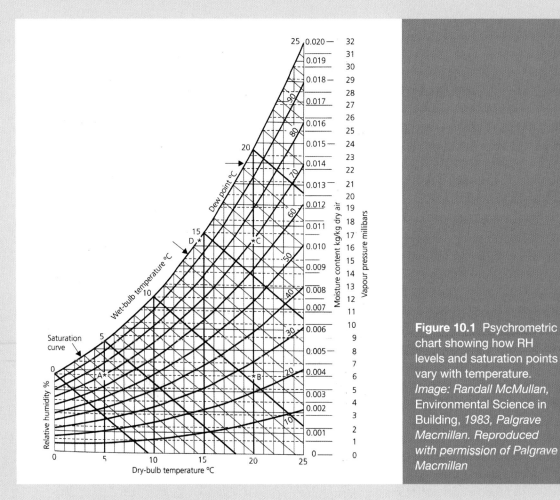

Figure 10.1 Psychrometric chart showing how RH levels and saturation points vary with temperature. *Image: Randall McMullan, Environmental Science in Building, 1983, Palgrave Macmillan. Reproduced with permission of Palgrave Macmillan*

potential of water vapour at that temperature – the air is effectively saturated. RH is a percentage calculation between the absolute humidity and the maximum absolute humidity; 100-per-cent RH would therefore mean you had reached the saturation point of the air at a given temperature, and condensation will occur.

When we devise the ventilation strategy for our buildings, we are aiming to achieve good RH levels, normally 30-60 per cent. Within this range, the beneficial aspects of water vapour in the air are optimised and the negative aspects minimised.

The amount of moisture in the indoor air is obviously influenced by the external climate. If your ventilation rates are high (lots of air changes), then the absolute moisture levels are likely to more closely follow the external air conditions. In order to control RH levels, it is important to keep ventilation rates low, as is the strategy in ultra-low-energy buildings. When we say 'low', this is of course only in the context of the general unplanned over-ventilation of existing houses. In ultra-low-energy houses, ventilation rates can be reduced to appropriate levels by using planned ventilation strategies and by minimising air leakage.

Relative humidity and perceived comfort levels

Regarding comfort levels, **ASHRAE** (the American Society of Heating, Refrigerating and Air-conditioning Engineers) has undertaken studies that indicate a general preference for RH levels of 20-50 per cent. ASHRAE is an international technical society that develops standards and guidelines on ventilation matters; these are normally adopted by, or inform, other more localised standards. ASHRAE considers an RH level of 35-40 per cent to be optimum during the heating season. Drier air will feel colder to us and we may also suffer from dry eyes or nose irritation, making us feel less comfortable. If the RH level is closer to our ideal, we will be happy with slightly lower air temperatures.

Relative humidity and health

The ASHRAE health standard for RH recommends 30-60 per cent for any habitable space, to minimise the growth of allergenic or pathogenic organisms. The 'Sterling Study', an ASHRAE technical paper,[2] looks at the effects of RH levels on a variety of unwelcome outcomes, such as dust mites, fungi, viruses, bacteria, respiratory infections, allergic rhinitis, etc. The results are shown in the Sterling bar graph (see Figure 10.2 below), which visually captures the increasing negative effects of either too-high or too-low RH levels.

There is increasing evidence to demonstrate the link between RH and these health-related outcomes; certainly enough for these effects to be widely acknowledged. With growing numbers

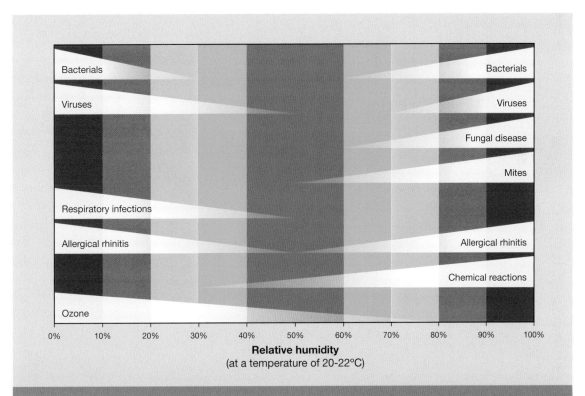

Figure 10.2 Sterling bar graph illustrating the link between relative humidity and health.
Source: Sterling et al. 'Criteria for human exposure to humidity in occupied buildings'[3]

of asthma sufferers since the 1970s (one in three Irish children has asthma), the need to ensure good ventilation and RH levels is key to the health quality of our new low-energy housing. Currently, indoor RH levels during winter are commonly below 40 per cent. If we can increase RH levels above this, we would reduce respiratory infections and asthmatic attacks in particular. If higher RH levels occur in summer, then there is an increase in the level of dust mites (almost eliminated in winter at lower RH levels, say, below 50 per cent). The number of people allergic to dust mites is also increasing, and therefore aiming to keep RH levels to lower levels, i.e. below 50 per cent even in summer, is very desirable. From current research it would seem that a target RH in the mid-to-upper 40s is the optimum range.

Relative humidity and the Passivhaus standard

The target RH level for the Passivhaus standard is 35-55 per cent. As we have seen from the above, aiming to keep the RH levels towards the middle of this range would seem to be ideal. Once you have achieved low ventilation rates, RH levels will be affected by the building construction itself and by how the building materials interact with the moisture in the air. Using materials to improve RH levels is therefore a strategy worth considering in an ultra-low-energy build.

The reasons for the Passivhaus adoption of a whole-house mechanical ventilation strategy are discussed in detail in Chapter 12, but it is worth noting that this ensures that moisture is being continuously removed directly to the outside, from the wet and moist areas of the house (kitchens/bathrooms). The overall effect on indoor air quality of RH levels (too dry or too humid indoor air) and the way this can be addressed is also discussed in Chapter 12.

Moisture in materials

Once we have ensured our building is airtight, and have therefore eliminated moisture transport into the fabric via the internal air, we then need to consider how moisture might enter our construction directly, through the materials we choose. The way in which moisture will affect and interact with a material is dependent on three different properties: capillarity, hygroscopicity and vapour permeability.

Capillarity relates to the uptake, release and accumulation of liquid water. **Hygroscopicity** relates to the uptake, release and accumulation of water vapour. **Vapour permeability** relates to the rate at which water vapour will be transported through a material, or its resistance to water vapour passing through it. These three properties of the materials you choose will affect whether moisture (both vapour and liquid) enters your walls, roofs and floors, how much will enter, whether it is then transported through the structure (and in which direction and at what rate!) or whether it is likely to accumulate at a certain location.

Capillarity

When water hits a building as rain, some bounces off, some will run down the face of the building and some may seep into the fabric. Liquid moisture can be relatively quickly transported some distance into the fabric, especially if the wall is constructed from a dry, porous material such as brick. This 'wicking' action is referred to as capillary action. Some materials have higher capillarity than others: gypsum plaster is a good example, as it is highly porous with myriad small pores and therefore has a very large internal surface area. Materials such steel or glass are not porous and have an effective zero capillarity. Driving rain is the main potential source of liquid water within any building and is therefore a very important aspect of water management to consider, especially when adopting less familiar building assemblies and if building on exposed sites.

Hygroscopicity

Moisture can also be taken up into materials hygroscopically, as vapour from the air. All materials have some level of hygroscopicity. Some will absorb vapour far more readily than others, and in the building industry the term 'hygroscopic' is generally used to denote a material that has either a high equilibrium moisture content (EMC), i.e. it can hold a lot of water vapour, or a fast rate of vapour absorption. (ECM refers to the moisture content of a material for a given temperature and humidity. When the ECM is reached, the material will no longer gain or lose moisture unless the ambient conditions change.) Hygroscopicity is generally considerably higher in purely organic materials than in purely inorganic building materials. The capacity of a porous material to carry liquid water through capillary action will normally far exceed any moisture taken up hygroscopically. This is certainly true of bricks or timber, for example.

A highly hygroscopic substance that can thereby be used to preserve a state of dryness in its local vicinity is called a **desiccant**. Rice, for example, is a fairly well-known desiccant and can be added to salt shakers to prevent the salt from becoming damp. Modern window spacers (which separate the individual panes of glass) will sometimes contain a desiccant to inhibit condensation between the panes.

Hygroscopic materials will naturally exchange moisture with the indoor air, if exposed to it, effectively flattening out ups and downs in RH levels – this can be a good reason to use certain organic materials (e.g. clay plasters). In an ultra-low-energy setting, anything that helps to avoid too-high or too-low RH levels is worth consideration. The key to such an effect is the rate at which the material will absorb and de-absorb (release) moisture. The capacity of a material for absorption of moisture will clearly also be related to its mass and to the exposed surface area available for moisture take-up. By intelligently selecting your construction assemblies, then, there is the potential to modulate RH and increase comfort levels.

It is worth noting that the cyclical wetting and drying of materials can have a detrimental impact, including encouraging general movement (expansion and shrinkage), especially in timber. Some materials' performance characteristics will be changed as they become moist – insulating performance especially can be compromised. Some natural insulation materials will retain thermal performance when wet, and this might be a very useful quality in some situations (sheep's wool is probably the best example of this).

Vapour permeability

All materials have a level of vapour permeability. In other words, some vapour will eventually diffuse through any material – whether natural, such as timber, or man-made, such as polythene. Given enough time and/or pressure differential, vapour will even diffuse through metal. However, this is unlikely to occur to any noteworthy degree in the course of a building's lifetime. Materials with poor vapour permeability are therefore described as 'vapour barriers' or, in the United States, 'vapour retarders'.

In the UK, materials that have a high vapour permeability are often described as 'vapour permeable', although this term can refer to a range of materials with differing vapour permeability levels. You may want to check what the actual vapour permeability is for a specific product. In the USA, materials can be defined as impermeable, semi-impermeable, semi-permeable and permeable, which is perhaps a more helpful means of distinction and provides a quick way to compare one material with another. Vapour permeability there is measured in grains of water vapour per hour per square foot per inch of mercury – known as US perms! (See Table 10.1 below.)

In the UK and Europe, four different units are used to measure the degree of vapour permeability of a material. To meaningfully compare materials, make sure that you have comparable units of measurement and that these relate to the thickness of material you are using. These differing and scientifically dense measures of vapour permeability may initially seem a little confusing, but you can convert between the four different measurements quite easily (see opposite) and then, simply, the lower the figure, the more permeable the material. The most important thing is to be able to compare the vapour permeability levels of the materials across your building assembly, which is not overly difficult to do.

The degree of vapour permeability in a material can be expressed in terms of the following. These values are summarised in Table 10.2 opposite.

- **Vapour resistivity (r-value)** – expresses vapour permeability relative to a standard metre depth of the material. The unit of measure is MNs/gm – meganewton seconds per gram metre.
- **Vapour resistance (G-value)** – effectively the same measure as the r-value but for a specific material thickness (MNs/g – meganewton seconds per gram). When assessing a material's vapour permeability within a structure, be sure to refer to its G-value – for example, emulsion paint might have a high r-value, but since you apply it in a layer only a few microns thick, its G-value will be relatively low. (Note that this unit is not to be confused with the 'g-value' that relates to window glazing – see Chapters 7 and 11.)
- **Water vapour resistance factor (μ-value, pronounced 'mu-value')** – a measure of a material's relative reluctance to let through water vapour in comparison with air. You would need to multiply this number by the material's thickness if you wanted to compare materials in your construction. As it is a relative measurement, it does not have any units. This value is commonly used in Europe.
- **Equivalent air layer thickness (Sd-value)** – expresses a material's vapour permeability as though it were a thickness of still air in metres. It is therefore related to a specific thickness of material.

Unfortunately, in the UK (at the time of writing), obtaining any of these values from manufacturers and suppliers can be difficult, partly because they are often not published along with the general technical properties. You can look up BS EN 12524, which includes the hygrothermal

Table 10.1 **US classifications of vapour permeability ('perms')**

Class	Permeability	Perms
Class 1	Vapour barrier / impermeable	≤ 0.1 perm
Class 2	Semi-impermeable	0.1-1.0 perm
Class 3	Semi-permeable	1-10 perms
No class	Permeable	≥ 10 perms

Table 10.2 **UK and European vapour permeability values**

Vapour permeability value	Units	Symbol	Thickness / relative to
Vapour resistivity	MNs/gm	r-value	One-metre thickness
Vapour resistance	MNs/g	G-value	Specified thickness
Water vapour resistance factor	No units	µ-value	Non-specified thickness, relative to air
Equivalent air layer thickness	metres	Sd-value	Specified thickness, relative to air

properties (those relating to moisture and heat) for many common materials. Since it is relatively easy to convert between the r-value, G-value, µ-value and Sd-value, figures can be compared even if you are only able to obtain, say, the r-value from one manufacturer and the Sd-value from another, as follows. (Unfortunately, perm ratings are not so easily interchangeable with the UK/European measures.)

- **To convert r-value to G-value:** multiply by thickness of material (in metres).
- **To convert G-value to r-value:** divide by thickness of material (in metres).
- **To convert r-value to µ-value:** multiply by 0.2gm/MNs (gram metres per meganewton seconds) – the UK value for the vapour permeability of still air.
- **To convert µ-value to r-value:** divide by 0.2gm/MNs.
- **To convert µ-value to Sd-value:** multiply by thickness (in metres).
- **To convert Sd-value to µ-value:** divide by thickness (in metres).

Table 10.3 overleaf gives the typical r-value and G-value for some commonly used construction materials, with the values for still air for comparison.

The vapour resistance of your materials is important in terms of how a particular wall or roof construction is built up, since it will determine how effectively any moisture can move through it (and out of it) and also in which direction it may travel. The potential for vapour diffusion across a construction will be related to the vapour pressure differential or gradient across it. As noted earlier in this chapter, for most of the year in a climate such as that of the UK, the vapour pressure outside (cool and dry) will normally be much lower than the vapour pressure inside (warm and moist), and if your construction is 'breathable', vapour diffusion will act to move moisture from the inside to the outside. The diffusion will occur evenly over the whole surface area of the construction, and the speed of diffusion will depend on how vapour-permeable the materials are. Generally, it is good practice to combine materials with similar moisture-managing characteristics. This is especially true for timber-frame construction.

It is useful to note that some materials can be quite vapour-permeable but not particularly hygroscopic. Mineral fibre insulation (e.g. stone wool) is vapour-open but not particularly hygroscopic and has limited capillarity – it will resist water take-up and will dry out relatively quickly if wet. Natural fibre insulations have some capillarity, high hygroscopicity and high vapour permeability. They will take up water easily – initially in vapour form, so will perform as if 'dry'. Once 'actually' wet with liquid water, they will dry out at a similar rate to mineral fibre.

Table 10.3 **Vapour resistivity and resistance values***

Material	Vapour resistivity (r-value, MNs/gm) for one-metre thickness	Vapour resistance (G-value, MNs/g)
Gypsum plasterboard Type 1	20	12mm board, G = 0.24
Still air	5	50mm air gap, G = 0.25
Mineral wool (rock or glass)	5	100mm thick, G = 0.5
Breather membrane		G = 0.1-2.0 (typical max. 0.6)
Gutex® woodfibre board	15	40mm thick, G = 0.6
Gypsum plaster	50	12mm, G = 0.6
Clay plaster	40	20mm, G = 0.8
Pavatex® woodfibre board	25	40mm thick, G = 1
Intelligent breather membrane		G = 2.0-50
Chipboard	100	22mm board, G = 2.2
OSB (oriented strand board)	200 (or much higher)	18mm board, G = 3.6
Brickwork	50	100mm brick, G = 5
Softwood	100	50mm-thick batten, G = 5
Hardwood	250	22mm oak board, G = 5.5
Plywood	450	18mm board, G = 8.1
Expanded polystyrene	300	100mm thick, G = 30
0.25mm polythene sheet		G = 250
0.12mm polythene sheet		G = 7,500
Roofing felt laid in bitumen		G = 1,000 (from BS 5250)
Glass		G = 100,000 (in the region of)
Metal		G = 100,000 (in the region of)
Vapour-control plasterboard		G = 100,000 (in the region of)

* Typical values only – variations occur.

High-performance 'closed-cell' insulations are not vapour-permeable or hygroscopic and do not have any capillarity. They will block moisture ingress but equally will block moisture egress. All three mechanisms (capillarity, hygroscopicity and vapour permeability) determine the potential moisture risks for any particular construction type.

Moisture management in construction

When considering a low-energy build, the two main sources of moisture within your construction will be via rainwater penetration externally and from moisture transported by the internal air. We do not cover the third source, groundwater, here, as this will be managed in a similar manner to conventional constructions – with a capillary break (see below and right).

Protecting your building from rain and other forms of precipitation is part of normal good building practice. The level of protection is of course affected by climate (in more aggressive climates, larger roof overhangs become more common), but also by the type of construction and its ability to manage moisture. Certain structures (e.g. brick or stone) will absorb liquid moisture and then release it in due course, and are referred to as a 'reservoir cladding'. Such structures will therefore experience less water run-off. A reservoir cladding will be capillary-open (i.e. subject to capillary action) and usually decoupled from the structural and thermal elements of the building, therefore this moisture should not adversely affect building performance. Alternatively, assemblies can be designed to protect the structure from any liquid moisture penetrating the surface (e.g. using cement render), called a 'drainage plain'. Drainage plains are either capillary-closed or have limited capillarity – they are effectively waterproof – and are therefore sometimes referred to as a capillary break. Since a drainage plain does not allow water to enter the structure, it will have increased water run-off. Reservoir cladding and drainage plains are the two main means of dealing with rainwater ingress in walls and roofs; both are designed to prevent rainwater reaching the structural and thermal elements of the building.

Your low-energy build should be essentially airtight, so moisture carried by the internal air should only be entering your construction materials via the three material transport mechanisms we have already outlined. Rainwater (and groundwater) transport moisture in greater volumes than any source of vapour moisture. However, vapour moisture transport can still create serious problems in a low-energy build due to the increased level of insulation, especially as the potential for interstitial condensation often increases (with multiple layers within the assembly) and the opportunities for moisture to escape or dry out can be reduced.

There will also be an element of inbuilt moisture in new constructions, resulting from wet trades and exposure to the elements during construction. Accidental or unforeseen moisture ingress, such as from leaky plumbing, can also be a significant source of moisture damage; even a 1mm gap in an airtight layer can allow significant volumes of water into the structure. Preventative measures do not usually involve matters relating to the fabric design and are therefore not covered in this book. However, ensuring that moisture can escape to the outside will be helpful. Even after a repair has taken place, saturated materials, if not replaced, may need time to dry out.

The challenge as regards low-energy design is that there are currently no standard 'robust details' available and no standard approaches, especially for retrofit projects. Making up unique details carries the risk of omitting to take 'something' into consideration – and at present we would suggest that moisture is perhaps the most likely 'something' to drop out of the low-energy equation. How your wall, roof or floor is managing moisture is something you need to think through carefully.

Breathability (or vapour permeability)

We refer to 'breathability' here because this is the term most commonly used, although we are essentially discussing the potential for vapour moisture movement.

In the first instance you will have the option to choose between using breathable and non-breathable materials for your assemblies. A breathable fabric will let more moisture into your assembly but will also let more moisture out, so will theoretically reduce the risk of interstitial moisture accumulation. Conversely, non-breathable fabrics will allow much less moisture in but, should moisture build-up occur, drying will take much longer or not occur at all.

Whichever your choice, it is advisable to maintain a level of moisture (and thermal) compatibility throughout – that is, use materials with relatively similar levels of capillarity, hygroscopicity and vapour permeability. An abrupt change in these properties can cause uneven distribution of moisture across an assembly, and may allow it to accumulate in some regions. If this accumulation of moisture occurs in a material with a high moisture tolerance (i.e. moisture does not significantly affect its performance or cause degradation), then it is of limited concern – assuming it will not accumulate to saturation point. If, however, it collects in an area containing moisture-sensitive materials, particularly if they comprise structural elements, you may have a serious problem.

Breathable assemblies

The rule of thumb to remember when designing a breathable assembly is the ratio 1:5. The vapour resistance of the material on the outside should ideally be five times less than that of the material on the inside. This excludes any service voids or rain screens. The moisture-driving potential is then set up to move from the inside to the outside of the assembly. From the values shown in Table 10.3 (page 154), then, we might choose to use woodfibre board (very vapour-permeable) towards the outside of a timber-frame wall and OSB on the interior side – this would deliver a sensible vapour resistance ratio. In this example, we would also be using the OSB as our airtightness layer and taping at all the board joints (see Chapter 9). While we are using a material that will allow some vapour movement, we do not want to allow any moisture carried by the interior air to travel into the assembly. The airtightness layer ensures this, and this is one reason it is placed on the interior side.

As already stated, ensuring that the airtightness layer is contiguous is important for both thermal performance and moisture management. How much moisture the air might carry into your structure through a non-contiguous airtight layer can only be roughly estimated. The photo opposite of a retrofit assembly shows moisture in a floor condensing against an internal brick wall that is colder than the surrounding structure. The moisture source is almost certainly from a combination of moisture held in the Leca® (clay-blown insulation), likely due to poor site storage prior to installation, and vapour moisture from the internal air – the airtight barrier was not yet taped to the wall. The problem should resolve itself, since both moisture sources are temporary, but it illustrates the need for materials to be dry (protected from rain, etc., during construction and stored appropriately). Check that materials are adequately dry before final finishing layers are applied. This is particularly relevant if using a product such as hempcrete (hemp plant mixed with a lime-based binder). Materials that are applied wet and then need to dry out over time must achieve appropriate RH levels before applying finishes that will inhibit or slow down further drying. The initial airtightness test (see Chapter 9) will identify any breaks in the airtight layer.

Non-breathable assemblies

While there are numerous benefits to designing your building to be breathable (mainly allowing water moisture to escape out of your assembly), for reasons of cost, time, space or planning, this may not be practicable. Indeed, the majority of high-performance insulation materials – such as extruded and expanded polystyrene or polyurethane and polyisocyanurate foam – have limited or (in the case of closed-cell insulations) no moisture-managing capabilities. If highly vapour-impermeable insulation is installed on the outside of a masonry wall, moisture contained in the masonry will take a long time to dry out (which means it's important to make sure the masonry is as dry as possible before applying it). However, bricks are relatively robust and will not deteriorate due to the effects of moisture in the same way that many other materials might. Greater risks occur if choosing to adopt a non-breathable material alongside more 'organic' materials such as timber. If choosing to use a vapour-impermeable membrane as your airtightness layer, this would need to be installed without error – any holes or rips could allow moisture in and the structure would then potentially struggle to dry out. Vapour-impermeable airtightness layers with timber-frame assemblies are commonly used in low-energy houses in cold European climates, although the level of experience in both timber-frame and using such membranes is much

Interstitial condensation forming at the junction of the floor-to-wall assembly in a retrofit. Dampness is visible at the base of the wall (below the plaster, where the skirting has been removed) and between the timber floor joists. The voids between the floor joists have been filled with a clay insulation (Leca®).

higher in those regions than in the UK. Unless you are very confident in your ability to detail such an assembly, in our view it is best avoided.

An example of a vapour-impermeable low-energy construction would be a rendered concrete structural wall that is externally insulated with a high-performance, vapour-closed insulating material (say polystyrene or phenolic foam). The materials' moisture-management characteristics are then well matched. Any risk of moisture entering the wall assembly would then be through cracking in the external render or poor detailing around openings, although with such materials (unlike with timber constructions) there is likely to be little movement to exacerbate such cracking. If the insulation were to experience prolonged exposure to moisture, there would be some material deterioration and thermal performance would be compromised, but if the rendered surface is maintained, the risk is small. Ideally you would detail around openings so that if water entered the assembly it could drain back to the outside. What this construction would not do is modulate the relative humidity (RH) of the interior air (effectively it has zero hygroscopicity). However, clay plasters could possibly be applied to some internal walls if some hygroscopicity was desired.

Modelling moisture levels

There are now useful tools for measuring hygrothermal (heat and moisture) transfer – in particular, for low-energy designs, dynamic numerical computer modelling (to BS EN 15026). Tools such as WUFI or Delphin can prove invaluable if you have an assembly where you are not completely confident that moisture will be managed safely. If you are unsure about a particular build-up of materials (especially in a retrofit), then such a modelling exercise may well be an excellent investment. Once a model is set up, you can alter materials within the assembly and, with such minor adjustments to your approach, you may well avoid what could have been an expensive mistake.

For measuring internal RH levels, you can purchase a hygrometer for under £10. At this early stage of adoption of ultra-low-energy design in the UK, gathering such data for wider application is very useful.

Example constructions

In this section we look at potential moisture issues that relate to some examples of ultra-low-energy construction. This cannot be an exhaustive list, but it does illustrate the principles already discussed, and should be helpful when applying the principles to alternative assemblies.

New build: cavity-wall construction fully filled with insulation

A cavity wall is a common example of a reservoir system. The outer leaf, often masonry, will soak up rainwater, but it is free to dry out on both sides. The sun and wind will help drive moisture through and out of the brick. The cavity acts as a separating layer and allows back-drying of the external leaf. In case moisture should penetrate the entire depth and gain access to the cavity, weep holes are built in to the brickwork of the outer leaf and waterproof cavity trays encourage moisture to escape to the outside.

Once you fully fill a cavity construction, you lose this back-drying capacity. Experience gained from fully filling traditional cavity walls suggests that using vapour-impermeable insulations on a site that is very exposed (say, a coastal location) can cause moisture problems, with water building up within the construction and deteriorated performance of the insulation. Some councils have stopped fully filling cavities for this reason.

To achieve ultra-low energy levels in a new build, you will be looking at cavity insulation of around 300mm. In order to ensure that your outer leaf is effectively managing water, it might be worth considering a rendered finish (a lime render on brick would reduce water absorption while still allowing some drying out), or you could protect the wall with an alternative drainage plain – tile hanging being another possibility. Deeper roof overhangs might be another helpful strategy (and will also provide summer shading).

The Certified Passivhaus building at Denby Dale in Yorkshire (pictured opposite) comprises stone external leaf, which is less capillary-active than a soft brick and allows use of a harder mortar joint – again limiting moisture ingress. The house also has a significant roof overhang.

If you want to retain a brick finish, perhaps select a less porous brick and then always ensure you maintain your mortar joints (the problem with this is that maintenance is not often our favourite activity!). You may also need to consider treating the external face of the bricks with a vapour-permeable capillary block (a transparent liquid coating that acts as a water repellent).

To get the required level of thermal performance, you may well need to use a mineral insulation in the cavity. This would be vapour-open but not hygroscopic, so if moisture reaches it, the water will drop under gravity. You must then ensure that any moisture at the wall's base is encouraged to drain to the outside. If the insulation did get moist periodically, it would affect thermal performance to some degree. If in a very exposed location, this may not be the wisest construction assembly.

The internal leaf could be a silicate block, which has good hygroscopicity and would help to modulate internal RH levels.

New build: timber-frame construction

In timber constructions, a rain screen is commonly used to protect the main wall assembly from the elements. The rain screen is often in

Denby Dale, Yorkshire: a masonry cavity-wall build constructed to Passivhaus standard.
Image: Morgan O'Driscoll Photography

timber and sits on battens (plus counter battens) in front of the main thermal structure. The screen protects the wall from most rainwater and can dry out from both front and rear (the rear must be a ventilated void). Such a rain screen is probably best described as a reservoir cladding system, since it absorbs and de-absorbs moisture cyclically.

If not using a rain screen, then your wall may be rendered or tiled (tiling is a form of rain screening). These renders can be watertight (a drainage plain) but still vapour-open, allowing some drying to the outside.

If using a timber-frame construction, choose insulations and finishes of similar character. Combining high-performance, diffusion-tight (vapour-closed) insulations with timber is not advisable as it could trap moisture and stop the assembly drying out, leaving the timber frame vulnerable to rot and mould. Disasters have occurred where rigid phenolic insulations have been externally applied to timber-frame constructions. This is a classic example of materials with different characteristics being paired together.

Apply the 1:5 rule (see page 156) when using vapour-permeable materials. In the summer months you can encourage some drying to the interior by using an 'intelligent' (humidity-variable), diffusion-open membrane as your airtight layer. This will generally become more vapour-open during the summer and add an extra potential drying route to the inside (see Figure 10.3 below). It is always essential that the airtightness layer is contiguous.

On both timber-frame walls and roofs, it may be that materials providing the usual racking strength (stiffening the timber frame) are in non-ideal locations for vapour permeability. It may be preferable for such materials to move to the inside of the frame, where they won't inhibit drying to the outside and where higher vapour-permeability levels can be tolerated. OSB could provide this racking function as well as forming the airtightness layer, for example.

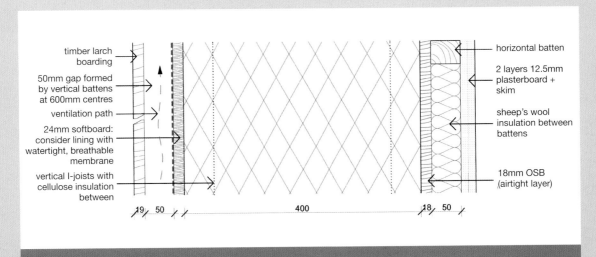

Figure 10.3 Section through a vapour-open timber-wall construction.

Retrofit: internally applied insulation with solid masonry wall

Unfortunately, this type of construction often introduces a number of points of moisture risk. If you are to insulate internally, be very careful when considering your moisture-managing strategy.

Existing solid-wall buildings – such as double-skin brick or stone-and-backfill – rely on the thickness of the wall and its high level of capillarity to manage moisture. Water will generally be driven into the wall only to a certain depth before a drying cycle will begin; in other words, there will be cyclical wetting and drying. Moisture will often dry out via the mortar joints and these will then deteriorate over time but can be repointed periodically. Repointing with cement mortars inhibits this process and can cause damage to soft bricks or stones, which end up retaining too much moisture and will break under cyclical freeze–thaw conditions (often the faces will 'pop' off). The first point to consider before internally insulating is therefore the condition of the external wall – is it likely to require some remedial work to make sure it is in a good state of repair?

Insulating internally will isolate the existing wall from the interior, including the heat that provides the driving force for drying in winter. If left cold and capillary-open, repeated rainfall will lead to greater accumulation of moisture, particularly if a non-breathable insulation is used. This can potentially lead to saturation of the existing structure, and if or when the temperature drops below freezing, this water will freeze, expand and widen cracks in your walls, leaving them more exposed to future moisture ingress. In dry summer conditions, south-facing elevations may become too dry, as they are no longer able to draw on moisture from the internal environment.

The Passivhaus Institut advises treating the external surface of masonry walls with a vapour-permeable capillary block when internally insulating. This will prevent rainwater ingress while still allowing drying out. However, studies have shown that the level of treatment required (recoating frequency) can exceed that recommended by the product manufacturers. Any such application will require regular maintenance to ensure it remains waterproof over time (recoated, say, every two to five years). It is also vital that this hygrophobic (water-repellent) treatment penetrates the brick or stone adequately to prevent localised moisture penetration at mortar joints. Before applying such a finish, you would also need to ensure that the quality of the existing wall is good – perhaps locally repointing mortar joints wherever they are eroded.

Rendering brick and stone walls (say, with a soft lime render) is one effective means of providing some weather protection from driving rain while still allowing the wall to dry out. Lime renders and washes are often referred to as 'sacrificial'; they weather-protect the stone or brick but wear out over time and are then reapplied – a very sensible and robust moisture-managing strategy. Lime mortars have some capillarity, so small amounts of moisture might still be drawn into the wall but will also be able to dry outwards.

In the Passivhaus-certified retrofit in Princedale Road, West London (pictured on page 77), the internal insulating wall was separated from the external solid masonry, creating a hybrid 'cavity' construction (see photo overleaf and Figure 10.4 on page 163). This approach treats the external solid wall as a reservoir cladding, but does mean that the brick wall is isolated from the warmth of the interior (which would normally help to drive moisture to the exterior). By introducing pathways for the moisture to escape from the back of the wall to the outside (using weep holes through the external brick skin), suitable conditions were created for some back-drying. We understand that the external brickwork was also treated with a capillary block. The material used for the framing out (the basic structure) of the cavity might be susceptible to periodic wetting and therefore needs to be selected appropriately – not timber. The internal insulation in this case was high-performance and vapour-impermeable, so there would be no drying to the interior or transfer of moisture through the materials from the interior air – as long as the airtightness barrier was not breached. Natural insulations such as sheep's wool or woodfibre are less thermally efficient than insulations such as phenolic foam boards, and therefore require additional depth for full Passivhaus Certification, resulting in excessive loss of floor area – which is why they were not used in this project. Even with high-performance insulation there is some floor-area loss; in the Princedale Road house this was partly compensated by the removal of chimney breasts – a moisture risk if retained. You would not want to ventilate these for energy reasons, and, if not ventilated, moisture build-up is inevitable. In this type of retrofit solution the depth of the internal insulation will change the overall proportion of the rooms, which may alter the aesthetics dramatically.

One reason why internally insulating a building to Passivhaus levels can be a great moisture-management challenge is that any thermal bridges will concentrate moisture condensation into localised areas. Making sure these bridges are addressed is then critical (see Chapter 8) and will be quite invasive. Avoiding penetration of the timber floor joists through your internal insulation is therefore important: possible details are shown in the photographs on page 139.

If retrofitting an existing building and your only option is internal insulation, you need to

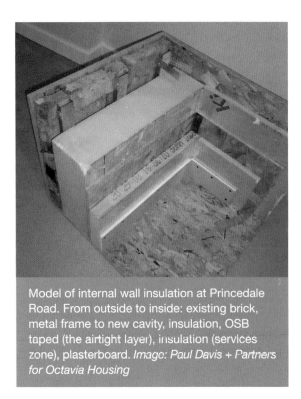

Model of internal wall insulation at Princedale Road. From outside to inside: existing brick, metal frame to new cavity, insulation, OSB taped (the airtight layer), insulation (services zone), plasterboard. *Image: Paul Davis + Partners for Octavia Housing*

consider carefully the level to which you can sensibly improve the building. For ultra-low energy you need to be willing to take on a radical retrofit project, aiming at giving your building its next 60-100 years of life, not just a 10-year refurbishment.

Retrofit: externally applied insulation

In terms of managing moisture, improving the thermal performance of existing buildings is best done externally. An external insulation system that will protect the existing structure from rainwater penetration can be easily designed. The existing structure then sits on the warm side of the construction and will be at much-reduced risk of water vapour condensing on to any of its surfaces. Thus the risk of condensation on the ends of timber joists penetrating the walls is virtually eliminated.

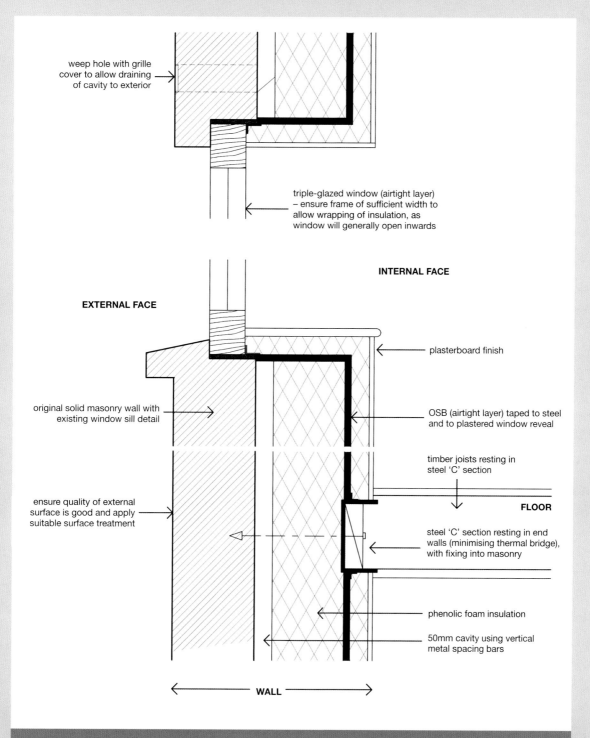

Figure 10.4 Retrofit section showing airtightness layer and internal insulation, following the principles applied at Princedale Road.

These types of insulation are generally referred to as either external thermal insulation composite systems (ETICS) or external insulation systems (EIS). The insulation is applied to the outside of the masonry construction, whether solid masonry or cavity wall (pictured on page 138), and finished with a render that protects it from rain penetration, acting as a drainage plain. Figure 10.5 below shows a section through such a wall. If externally insulating a cavity construction, you should also fully fill the cavity to avoid thermal bypass. Try to ensure it is fully filled without air gaps – expanded polystyrene-graphite-coated beads (InstaBead) are good in this respect, as they flow well.

External insulation will normally mean having to adjust your roof and wall junction to suit the new wall depth, unless there are already deep roof overhangs.

Fixings or small cracks in the finishes will introduce potential weak points through which driven rainwater might enter. Any moisture drawn in at these points by a capillary-active material behind may struggle to get out again if the drainage plain is vapour-closed. Similarly, if a vapour-impermeable external final finish (some paints) is applied, this may also inhibit drying. You should generally assume that some moisture will get in behind the rendered surface;

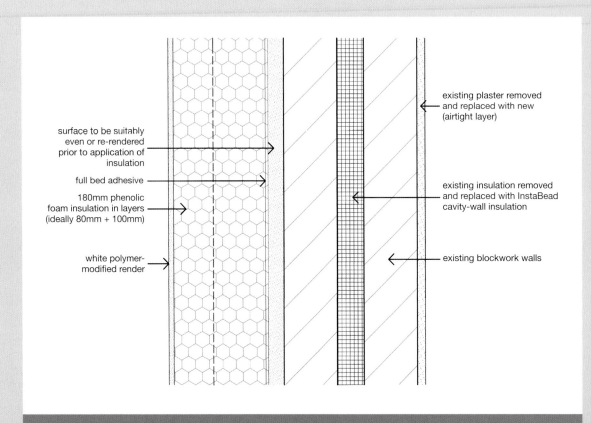

Figure 10.5 Section through an externally insulated cavity-wall construction at the Totnes Passivhaus.

thus a level of vapour permeability in the drainage plain is advisable. (As we have seen, a material can be watertight but still retain some vapour permeability.) Generally, thin render systems are used, either polymer cement or silicone- or mineral-based renders, all of which allow for some transmission of water vapour. When the sun warms such a rendered wall, some drying out can then take place.

ETICS generally use high-performance insulations such as rigid phenolic foam boards, which have no capillarity, hygroscopicity or vapour permeability. When this type of system is applied to a rendered masonry wall, the possibility of any serious moisture damage is limited. There are no 'organic' materials to rot and the insulation is moisture-resistant. An appropriately specified cement-based render, which can still provide some vapour permeability, is therefore suitable for use with such insulations. The key is to maintain the render surface and to detail well around windows and doors to avoid undue water ingress – wind can blow rain some distance up cracks and gaps. The detrimental effect of water lying temporarily against such insulations will be minimal. Far more problematic in this scenario would be if air could enter and then circulate behind the insulation, compromising its performance (see Chapters 8 and 9).

Alternatively, a more 'natural' ETICS solution would be to use woodfibre boards as the insulation material and then protect this with a lime render, which has a much higher permeability than cement render – again, joining materials with similar characteristics. Any moisture that gains access will get absorbed by the woodfibre boards, which are relatively hygroscopic. With a good vapour-permeable lime-based render (which may have some capillarity as well), the moisture should be able to dry out cyclically. Lime renders also have the ability to self-heal – sealing up small cracks over time. Obviously, the ideal is to avoid any moisture penetration, but this approach allows for the moisture to be managed, limiting any potential negative impacts. It is important to note that the characteristics of lime plasters and renders can vary enormously, so make sure the recommended lime mix matches the insulation product being used.

A woodfibre board will perform less well thermally than a phenolic foam insulation, which means thicker walls and greater loading on the existing structure. Another approach, then, might be to create an external timber frame with a thin woodfibre board cover, which could then be filled with a lighter natural insulation, such as cellulose. This would reduce the loading on the existing structure. You would need to match the materials in the assembly according to their moisture-management characteristics. The problem with external insulation is that it will not always be an acceptable solution aesthetically, especially in conservation areas, so upgrading a property may be possible only by insulating internally.

Flat roofs, including living roofs

It is worth commenting on flat roofs, which require vapour-impermeable finishes as your drainage plain (asphalt, felts, etc.). If you are using breathable materials, such as timber, you need to maintain vapour permeability right to the outside. The way to approach this is to use the 1:5 principle (see page 156) and introduce a ventilated void between the watertight layer and the top of the main insulated roof assembly. This means that the watertight layer works in a similar way to a rain screen on a vertical wall. If the roof is of any great size (or you have roof lights penetrating the void), you need to be confident that air is able to freely ventilate the entire void area. Blocking drying to the outside is not a clever idea!

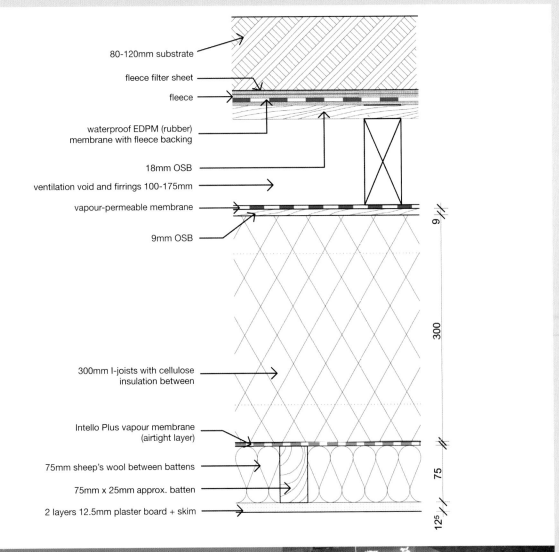

Figure 10.6 Section through living roof 'vapour-permeable' construction above main timber roof utilising a ventilated void at Totnes Passivhaus. The photograph on the right shows how it was executed on-site.

For timber flat roof constructions (see Figure 10.6 opposite), an intelligent membrane on the inside of the assembly as the airtightness layer might be of benefit. The flat roof needs to be covered with a dark-coloured material (such as asphalt), which will then absorb heat from the summer sun – dark surfaces will absorb significantly more heat energy than light ones. This warm external structure should then drive any moisture trapped within the assembly towards the inside, allowing potential drying to the interior. Moisture will always be driven away from warmer areas towards colder areas. However, this does not obviate the need for this type of assembly to have a ventilated void.

Boards providing racking strength (stiffening the timber frame) and with high vapour-permeability levels should, if feasible, be located towards the inside of the roof assembly so as not to unduly inhibit drying to the outside. With a timber-assembly living roof where there is a ventilated void, the board above the insulation could be eliminated altogether and replaced solely with a vapour-permeable membrane.

Concrete flat roofs

Where the roof assembly uses non-breathable materials, e.g. in the case of a concrete roof deck, it is possible to insulate externally using a product such as Foamglas® or other similar closed-cell-structure and zero-vapour-permeability insulations. This keeps similar-acting materials together and there is no need for a ventilated void.

RECAP

The consideration of moisture management within and through highly insulated assemblies is an important aspect of achieving healthy and durable buildings. It is the potentially detrimental effect of moisture on indoor air quality (in relation to relative humidity and mould), as well as the risk of material degradation, that make this so vital to an ultra-low-energy build. The integrated approach to moisture management and low energy performance is an illustration of how the principles of Passivhaus need to be adopted as a whole.

Moisture is transported in both liquid and vapour forms, and potential water ingress into the structure needs attention both from the external and internal climates. Understanding the mechanisms of moisture transport (capillarity, hygroscopicity and vapour permeability) is key, and both the airtight layer (to the inside of the assembly) and the wind-tight layer (to the outside of the assembly), play important roles in this regard. Encouraging water vapour to move towards the external side of your construction is best practice, as is combining materials with similar characteristics.

In retrofit situations, the management of moisture is even more critical, especially if insulating internally. If carrying out an unusual detail where there may be a risk of condensation (especially if within the assembly itself, i.e. interstitial condensation), then modelling constructions in dynamic computer programmes such as WUFI makes good sense.

CHAPTER ELEVEN
Windows

High-performance windows, U-values (frame, pane, spacer and installation factors), solar gain, window installation, window sills, doors, roof lights, avoiding summer overheating, the construction phase, future developments and costs

Windows contribute a wide range of functions to any building, from the aesthetic to the purely practical – natural lighting, a visual connection to the outside, ventilation, sunlight, sound insulation, and so forth. In a Passivhaus or other ultra-low-energy design, windows play an important role in the overall performance of the building, both in terms of comfort and energy in use. This is why the measured thermal performance of the windows is part of the required standard: it's not simply an optional extra. Even with high-performance windows, the heat/energy losses through them will be around five times more than losses through the equivalent area of walls, roof and floor. Furthermore, passive solar gain through the windows makes a significant contribution to the building's overall heat gains. Optimising window performance is therefore key to ultra-low-energy design.

Development of high-performance windows

Window design has already developed from single-glazed to double-glazed units. Originally the gap between the two panes was filled with air, but now this is commonly filled with argon, as it has a lower thermal conductivity than air. Leakage of the argon has been a problem with regard to retaining long-term performance, and, while lower leakage levels are now achieved, performance will drop as the argon dissipates (see Chapter 7, page 99). With currently available products, most of the gas may be lost within the first ten years of the window's life, although product performance is improving all the time. Other gases (krypton and xenon) are also sometimes used; these are significantly more expensive but further improve thermal performance. Krypton and xenon are relatively rare, and utilising them for windows is not encouraged

by the PHI. This resistance to their use is also related to the common use of thinner spacers between the panes of glass (10-12mm), which means that once the gas has leaked, performance drops much more markedly.

In addition to filling the gap with gas, one of the surfaces of the glass is now commonly treated with what is termed a 'low-emissivity' or 'low-e' coating – an extremely thin metal-oxide layer. All materials reflect or absorb radiant energy, but to very different degrees. Black materials, for example, have high thermal emissivity – they will absorb high levels of radiant energy and therefore become very hot in the sun. Materials with low thermal emissivity have high reflectivity. The effect of the low-emissivity layer is to reflect long-wave radiant heat from indoors back into the house, thus keeping the house warmer in winter. Short-wave radiation from the sun is still able to pass through the coating (although it will also reflect this back to the outside to a small degree). There are hard and soft low-e coatings; the hard coating being more durable but less efficient. A standard triple-glazed window unit will be argon-filled and have two low-e coatings, on the inside surface of the inner and outer panes, which are thereby protected from scratching, etc.

In future it may be that the glass panes will be separated by a vacuum, using nearly-transparent pillars or spacers. This will improve performance even further but is not yet commercially viable. One of the challenges will be to retain the vacuum over the window's design lifetime.

Many low-energy windows open inwards, which is unfamiliar in the UK. This is partly owing to the need to insulate around the frame externally, which makes opening outwards problematic. Inward opening can make it somewhat impractical to display items on your window sill, although this problem is avoided

A tilt-and-turn inward-opening Passivhaus-suitable window. With the window open in tilt mode, the window ledge can still be used.

with tilt-and-turn windows, which also offer security and rain protection advantages when tilted (see photo above). Some manufacturers do offer low-energy externally opening windows.

U-values

The thermal performance of a window is often given by a single U-value. You can ask your supplier for this value, even if it does not appear on the 'advertising' packaging. In the UK it is still common to appreciate only that there is a performance difference between single-, double- or triple-glazed windows (see Table 11.1 below). However, double- or triple-glazed units can perform very differently, with a wide range of U-values. The U-value is a function of the thermal conductivities of the different materials making up the window and their thicknesses: the lower the U-value, the better the thermal performance. Manufacturers may quote the U-value as calculated through the centre of the window pane, since this gives the most favour-

Table 11.1 **Typical domestic window U-values in the UK**

Window	U-value (W/m²K)
Single-glazed	4.0-5.0
Typical UK double-glazed 2010 building stock	approx. 2.0-3.0
New-build double-glazed, compliant with current (2010) Building Regulations (England and Wales)	<2.0
Passivhaus (triple-glazed)	<0.80

able results. In fact, the thermal performance of a window is affected by a much wider variety of factors, which are discussed here.

Typical window performance U-value: >1.5W/m²K

The worst U-value you can achieve with a construction is approximately 5.0W/m²K (watts per square metre per degree kelvin). This is for a single-glazed window – which is why changing from single to double glazing makes such a huge difference to energy performance and comfort levels! The poor performance of windows means that the internal surface temperature of the glass will normally be the lowest in the house (older readers will probably have memories of scraping ice off the inside of window panes on cold winter days). The lower the internal surface temperature of the glass becomes, the more uncomfortable it is for the building's occupants, because it creates an asymmetry with other, warmer wall surfaces. We feel the contrasting radiant temperatures from these different surfaces and don't like it! This is one reason why radiators are commonly placed under windows – to compensate for this effect. If surface temperature variations are greater than 4°C, we will feel the difference and choose not to sit next to windows during cold periods, effectively reducing the usable floor area of our homes. Furthermore, the cold surface of the glass cools the air adjacent to it and the air then sinks down, spreading out across the floor and setting up air rotation within the room. This can have the further effect of reducing temperatures at lower levels (around your feet) and increasing temperatures higher up (at the ceiling and around your head). If this temperature stratification becomes greater than 2°C, again it starts to feel uncomfortable.

The current (2010) Building Regulations in England and Wales stipulate a maximum U-value of 2.0W/m²K for windows in new dwellings.

Passivhaus window performance U-value: <0.80W/m²K

A Passivhaus-certified window must achieve a whole-window U-value (see below) of 0.80W/m²K (in a cool-temperate climate). Part of the reason for this particular performance level is to ensure that the internal surface temperature of the window is high enough to keep variations of surface temperature within any room below 4°C. This removes the need for a heat source adjacent to the window, and also means you can remain comfortable wherever you choose to sit, effectively maximising usable floor area. This surface temperature of the internal glass will be high enough to ensure that you don't get temperature stratification in the room, which again is likely to boost comfort levels.

Different climatic regions will require different target window U-values and different glazing types to meet the Passivhaus standard. Full details of these targets, relating to seven different climate regions (from 'arctic' to 'extremely hot, often humid'!), are published by the PHI,[1] along with examples of cities within each region. In a warm climate (e.g. Hawaii), double-glazed units are specified, whereas in a cold climate (e.g. Anchorage), a quadruple-glazed low-e window is required.

Calculation of the window U-value

Although in the UK you may still be quoted just a single U-value for a given window, taken through the centre of the pane, manufacturers here are now beginning to quote a more complex whole-window U-value, U_w (see box overleaf). However, this differs from the whole-window *installed* U-value that is calculated by the PHPP (see page 174). The three different factors taken into account for this are as follows.

1. Differences between the window frame and the window pane

In reality, every window is made up of two very different 'constructions': the window pane and the window frame. The single U-value calculation can therefore be very misleading, especially if it is based on the window pane only, since this will perform much better than the frame. The window frame will generally be the weakest point in the performance of a building's thermal envelope. The construction and design of the frame is thus a key focus point for manufacturers of high-performance, triple-glazed windows – commonly the frame will be made of multi-layered bonded wood, with some insulation. This could be hard or soft polyurethane such as purenit® (hard), extruded polystyrene, cork or cellulose. The whole-window calculation takes into account the following:

- The U-value as calculated through the centre of the window pane (U_g).
- The U-value as calculated through the window frame (U_f).

(The above two values are adjusted to take into account the proportion of glass to frame.)

- The window spacer psi-value (explained on page 174).

The whole-window U-value will therefore change according to the proportion of frame to glass. Since the frame is the weaker component, a larger window will perform more efficiently than a small window. Additionally, a window frame designed to be openable will perform worse than a fixed, i.e. non-openable, window frame. So, again, the straightforward conclusion is that large, fixed glazing will perform best of all (as well as being cheaper!). This simple understanding can be very helpful at the design stage – installing a fixed window on a north-facing façade is a good trick to minimise heat loss while still maximising views and light. Not all windows need to be openable, as long as there is one in each room.

Section through a typical Passivhaus-suitable triple-glazed window, in this case with an aluminium finish externally and wood internally.

> **Whole-window U-values as quoted by window manufacturers in the EU**
>
> When a U-value for the whole window is quoted (U_w), it is for an assumed standard size and type of window: 1230mm high by 1480mm wide, divided by a centre mullion, with one fixed and one opening casement. This helps when making comparisons between different manufacturers.

The U-value of the window frame in Passivhaus-suitable windows can vary by around 0.13W/m²K, and it has been shown that this could make a difference of more than 2.5kWh/m².a to the building's annual heat demand. On a certified project, this could easily make the difference between a pass and a fail. On a non-certified project, this may not be as critical as the cost savings that might otherwise be achieved – a slightly less efficient window frame can still perform as required. For Passivhaus Certification of a window, the frame U-value should be below 0.80W/m²K, and only a few manufacturers to date have achieved this.

2. The effect of the window spacer

The window spacer is the small edge piece that keeps the panes of glass apart. In effect, each spacer acts as a mini thermal bridge. In the UK, these spacers are mostly made of aluminium – a rather effective conductor! The fact that the spacer performs so poorly is one reason you tend to get condensation forming at the glass perimeter of standard windows, where the internal surface temperature ends up being the coldest, and this then encourages mould growth. You will never have a solid aluminium spacer in a high-performance window, but there are several alternatives that will be suitable, generally referred to as 'warm-edge' spacers. Examples include a thin-walled stainless steel 'U' channel, which is normally filled with a highly insulating plastic desiccant and has a separate moisture-proof butyl seal, or thermoplastic spacers, which are made from a semi-rigid plastic material with a desiccant entrained within them (the best-performing products to date). You can boost window performance by specifying the most efficient spacer, for obvious reasons. Durability may be an issue with some of these solutions, in terms of ensuring that the spacer does not allow any undue leakage of the argon gas, although windows can now achieve extremely low leakage levels. Also, new and

Silver-coloured warm-edge spacers in a Passivhaus-suitable window.

better spacers will become available as the market for high-performance windows grows.

The psi-value for the spacer is multiplied by the length of the spacer around the perimeter of each window to further adjust the final U-value calculation. All these calculations are done within the PHPP (see Chapter 7).

3. The installation of the window within the wall construction

It is now understood that the manner in which the window is installed in the wall will have a significant impact on its final performance. Investment in an expensive high-performance window could be wasted if the window is then not installed appropriately. A poorly installed window will create a significant linear thermal bridge (see Chapter 8) around the four sides of its frame, which will effectively increase the U-value of the component, worsening its performance. In fact, an extremely poor installation (similar to how windows are commonly and currently installed in the UK) will badly compromise your annual space heating demand on an ultra-low-energy build. For this reason Passivhaus sets a second window U-value requirement – a whole-window installed U-value ($U_{w,\ installed}$) of less than 0.85W/m²K. This is calculated within the PHPP.

A linear thermal bridge is calculated in W/mK, which is referred to as the psi-value (ψ). The psi-value for a construction detail is normally measured using a calculation package such as THERM (see Chapter 8, page 120). This value is then multiplied by the length of the thermal bridge to ascertain the final effect on the thermal performance of any particular window installation. In order to make this calculation, the construction detail for how the window is to be installed in the wall is critical, and (as we saw in Chapter 7) the psi-value of the installation detail ($\psi_{installation}$) needs to be entered into the PHPP (along with other values relating to the performance of the window itself). The designer's aim is to detail the junction around the perimeter of the windows to try to achieve a thermal-bridge-free solution. Within the IBO book (see Resources) there is a range of illustrated construction details. These are rather diagrammatic and also use some materials that are more familiar in Austria and Germany than in the UK; however, they can be extremely useful in developing appropriate and effective new-build solutions. The psi-value is pre-calculated for each of these details, which is ideal for the PHPP and at least gives a guide figure to follow at the initial design stage. There are two psi-values given – one for the head and sides of the window installation, and a separate one for the sill detail. We will look at sill detail later in this chapter, but suffice to say this is more difficult to detail for good performance.

With a really good window installation detail, you will improve the overall performance of the window. To achieve this, you will have thoroughly wrapped the window frame with the wall insulation (see 'Effective installation of windows', page 176). There are also one or two verified window details (with a PHI certificate), where manufacturers and installers have collaborated to optimise installation solutions and provide the psi-value for you – hopefully this trend will continue.

It will not be time-efficient to individually model numerous psi-values for a given project, so the best approach is to check against other similar details and be on the cautious side when determining a psi-value to be entered into the PHPP (perhaps discuss with the Passivhaus Designer or Certifier). It is only if you want a certified building and you are near the 'magic' 15kWh/m².a space heat demand that every little extra improvement might be needed. Then your Certifier may wish to see a full calculation for your window installation detail. In reality, for most designers it is an understanding of the principles of a good installation that is essential. Optimising the installation includes looking at reducing the time, effort and materials involved.

Calculating the whole-window installed U-value

Figure 11.1 opposite shows the formula for calculating the whole-window installed U-value. A calculated example is given here, based on the window elevation in Figure 11.2. U-values for the window glass (e.g. 0.70W/m²K) and frame (e.g. 0.80W/m²K) should be provided by manufacturers of Passivhaus-suitable windows, along with the psi-value for the warm-edge spacer to be used (e.g. 0.03W/mK – you must aim for a value below 0.05W/mK).

The psi-value for your installation depends on how good your construction detail is for building into your wall. If you use a suggested detail from a supplier or one illustrated in the IBO book, then a figure may well be given there, or you need to add a conservative figure of 0.04W/mK,

$$U_{w, installed} = \frac{(U_{glass} \times A_{glass}) + (U_{frame} \times A_{frame}) + (\Psi_{spacer} \times L_{glass}) + (\Psi_{installation} \times L_{installation})}{A_{window}}$$

Figure 11.1 U-value for a whole-window installation.

or calculate for your specific detail using a modelling package such as THERM. With a good detail, your psi-value may be 0.004W/mK, but for this example let's say your value is 0.20W/mK, so your installation detail is far from ideal. The units of areas and lengths are metres and metres squared.

So for the window in Figure 11.2, the whole-window installed U-value is:

= [(0.7 x 1) + (0.8 x 0.44) + (0.03 x 4) + (0.2 x 4.8)] / 1.44
= (0.7 + 0.35 + 0.12 + 0.96) / 1.44
= 2.13/1.44
= 1.48W/m²K

This is not a good outcome for a Passivhaus-suitable window installation, where you are aiming for a value of below 0.85W/m²K. If we assume the installation psi-value was 0.004W/mK

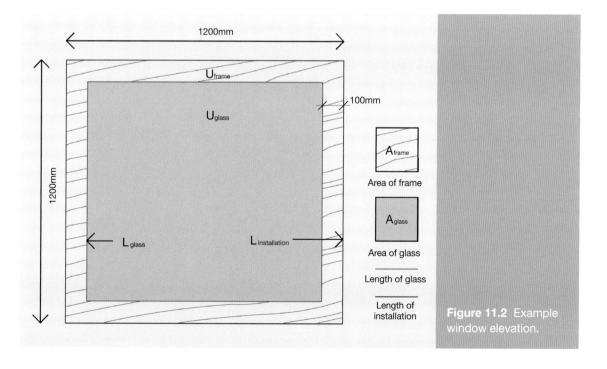

Figure 11.2 Example window elevation.

(a good install detail) the revised U-value calculation becomes:

= (0.7 + 0.35 + 0.12 + [0.004 x 4.8]) / 1.44
= 0.83W/m²K

The importance of a good installation is clear!

Solar gain

A Passivhaus gets its name partly from the 'passive' energy it will gain from the sun during the winter months. Capturing the winter sun has always been part of the low-energy approach, and exploiting any 'free' energy makes absolute sense. Passive solar gain will then make up part of the total energy balance (total energy losses balanced against total energy gains – see Chapter 6, page 76). For this particular function in an ultra-low-energy build, the glass needs to meet particular criteria. Firstly, the glass U-value (U_g-value) should ideally be <0.75W/m²K. Secondly, the glass should ideally have a 'g-value' (see below) of more than 0.5. In essence, the glass should minimise heat loss to the outside while allowing some sun radiation admission.

Solar transmittance and the g-value

The g-value is the fraction of the heat from the sun that enters through window glazing and is expressed as a fraction between 0 and 1. A low g-value means that the window transmits less of the sun's heat. A g-value of 0.5 is sometimes expressed as 50 per cent. In the USA, this figure is referred to as the solar heat gain coefficient (glazing), or 'SHGC-glazing'. Note that low-emissivity (low-e) coatings and argon filling both lower the g-value slightly.

In the northern hemisphere, south-facing façades will capture the most energy, while east-facing will capture the morning sun and west-facing the evening sun. For Passivhaus energy modelling, the g-value of all the windows is inserted into the PHPP, which uses an equation (including adjusting for building orientation, latitude, window orientation and shading) to calculate the windows' contribution to the overall energy balance. The PHPP does not overestimate what the passive solar gain from windows will be, which is another reason why it gives realistic energy 'in use' predictions. The window g-values are entered in the WinType worksheet within the PHPP.

In hotter climates, where wintertime solar gain is not needed and internal overheating is a potential issue, you will want a low g-value; a value below 0.35 would be preferable. The PHPP recommendation of a g-value of more than 50 per cent suits a cool-temperate climate, but will not always be appropriate.

Effective installation of windows

The four basic principles of a good installation are to make sure that:

- the window sits in the same line as the insulation, not staggered to it (the window forms part of the continuous insulation layer around the building)
- you wrap the outside of your window frame with at least 60-70mm of insulation
- your airtightness layer transfers from the wall to the window with an effective transition detail (normally using a tape on to the window frame)
- the window is wind-tight along outside edges and waterproof against driving rain.

Understanding these principles, which are not complex, will enable you to work out your own effective installation solution, which should be considered early in the design process.

The fact that in an ultra-low-energy build the frames are normally wrapped with external insulation means that only a small depth of the perimeter frame is visible on completion, allowing an almost frameless appearance. Most manufacturers offer timber window frames with an aluminium external facing for low maintenance. Some cut this metal facing right back (to, say, only 15mm or 20mm depth), exposing the insulation/wood frame (pictured below). This means that after fitting there is minimal aluminium facing embedded in the wall insulation. This is an effective low-energy strategy as otherwise the aluminium will act as a rather effective conductor and make the installation perform less well; it also saves on high-embodied energy material.

The following examples – typical window installation details for retrofit masonry and new-build timber-frame constructions – illustrate the principles of a good window installation.

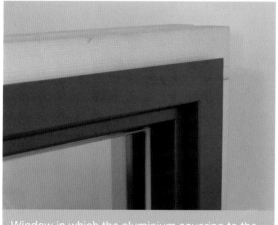

Window in which the aluminium covering to the external frame is minimised. It is designed so that the wooden 'uncovered' frame and part of the aluminium is wrapped by the wall insulation.

The significance of good window installation

If you decide to use conventional installation methods, this will decimate your window U-value! It will be an enormous waste of your hard-earned cash. Only buy ultra-low-energy windows if you can install them effectively. Conversely, for those not aiming for ultra-low-energy performance, you can boost your more average window performance with a really good installation!

Masonry wall with external insulation

This detail would be fairly common on a retrofit project, and of course, as we saw in Chapter 10, it is preferable to insulate externally rather than internally, for the following reasons:

- It avoids any potential moisture problems generated by internal insulation.
- It avoids losing internal floor area.
- The thermal mass of the masonry wall ends up on the warm side of the construction, which might help with summer cooling.

Make sure you pre-plaster ('parge') your masonry window reveals, prior to fitting windows, with a gypsum undercoat plaster such as Thistle Hardwall®. Block, brick and mortar are not airtight.

The window has to be fitted on the outside of the masonry wall so that it sits within the insulation zone. This involves structurally supporting the window with either steel angle

Window installation in a masonry wall.

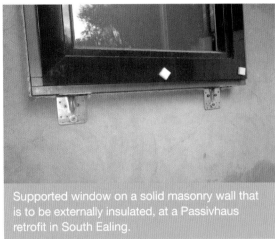

Supported window on a solid masonry wall that is to be externally insulated, at a Passivhaus retrofit in South Ealing.

brackets or a combination of metal straps/ties and a timber batten. If the window is conventionally sized, you should not need to resort to steel; the timber and straps should be sufficient. The result will be deep window reveals internally.

For airtightness you may tape on to the window frame, either on the internal face or on the frame sides (see Chapter 9, page 140). If taping on to the window frame face, aim to keep the tape width to about 10mm (there are tapes with multiple release papers for ease of application), otherwise you might have difficulty covering up the tape at the finishing stage; be sure to leave enough frame visible to allow your openable windows to swing inwards. The tape must then be embedded within the plaster coats.

If you are undertaking a retrofit project and improving the building fabric over a period of time, it would be worth considering installing new windows in the correct position with a temporary cover frame, ready for future installation of external wall insulation.

Lightweight timber 'I' wall

An 'I' wall refers to a timber-frame wall made up of I-shaped timber columns, i.e. two small section timbers joined together by a relatively thin web of timber. These are commonly used as the structural frame for walls and roofs.

In this case, the window sits within the wall in line with the insulation. If the main wall is insulated with a blown-in insulate such as recycled waste paper (cellulose), then the frame wrapping will need to be a more rigid material, such as a woodfibre compressed board. (Insulations such as recycled newspaper have a loose, 'fluffy' appearance and are often blown into voids through flexible pipes, the advantage being that the material tends to fill every nook and cranny and there is no waste.)

Airtightness can be relatively straightforward – in this example, it would involve taping from the window frame on to an internal oriented strand board (OSB) lining.

Window installation in a new timber wall.

Standard-profile sill used within external insulation.

Window sills

Windows do not usually come with integral sills, and these sometimes have to be sourced and ordered separately. One or two window manufacturers offer sills (see www.passivhaushandbook.com for details) as an additional service, which keeps responsibility and coordination to one supplier. There are pre-insulated hollow metal sills available from Europe, and your window supplier should be able to supply brochures and contact details for these products.

We found that getting information on these products in English was problematic, so in the Totnes Passivhaus we used a locally made standard-profile metal sill and simply ensured that the wall insulation ran right up to the underside of the sill profile. This proved to be as effective as a purpose-made sill, although it required some care and attention to detail on-site.

To avoid ingress of wind-driven rain, you need to fill any gaps around the perimeter of your sill with an appropriate waterproof sealant or with weather-tight flashing tapes. There is always some potential for water to be drawn in under the sill, so allow for any water to drain or dry to the outside.

Doors

Most Passivhaus window manufacturers supply a range of Passivhaus-suitable doors. Currently, Passivhaus-certified main entrance doors are sourced from only three European manufacturers, but are often supplied through the window manufacturers as well. The supply and installation issues are the same as with windows. The designs for front entrance doors tend to be uncompromisingly contemporary, and the doors themselves are surprisingly heavy (make sure you have fixed them securely in place before you slam the door shut!)

Passivhaus-certified doors are not cheap, although there is no requirement to use them – you can have a certified project without one. If budgets won't stretch to a Passivhaus-certified entrance door, there are also balcony doors, tilt-and-slide doors, patio doors, etc., based on standard windows, i.e. they are in effect fully opening windows. If the head, sides and sill are the same as a window, they can then be certified components. These might require you to step over the sill, as it may not be designed to stand on. Where you need the sill to be weight-bearing, you can usually insert a threshold detail – either a thermally broken glass-reinforced plastic threshold (insulated) or an ultra-low (i.e. shallow), thermally broken rigid PVC threshold for wheelchair access. The window U-value will need to be adjusted to reflect the final threshold detail. Some manufacturers supply units with

Insulated sill under a sliding door unit, at the South Ealing Passivhaus retrofit.

integral load-bearing insulated sills (pictured above).

If the installation is not dimensionally accurate (i.e. not skewed), you may encounter airtightness issues with opening units and seals.

Roof lights

If introducing roof lights, it is important to first consider the positives and negatives. One problem with roof lights is that they tend to sit proud of the main construction, breaking the good-installation rule number one (the window should sit in the same line as the insulation, not staggered to it) and making it difficult to meet rule number two (wrap the outside of your window frame with insulation). The centre pane U-value may be excellent but the frame U-value is unlikely to be great, and installation psi-values will almost certainly be poor. There is also an overheating consideration – roof lights do let in more light than vertical windows and therefore also more heat energy. However, if opened on hot days they could form part of a natural ventilation strategy (e.g. located at the top of

A tilt-and-slide Passivhaus-suitable door.

a stairwell – a passive stack effect), contributing to summer cooling. There are FAKRO® and VELUX® triple-glazed low-energy versions, some even with external shading devices to deal with the overheating risk.

Of course, not every detail in a Passivhaus build needs to be thermal-bridge-free, and you can choose to compensate by over-performance elsewhere. Generally we would currently advise designing out the need for roof lights, or keeping them to a minimum. However, improved products are already beginning to ease this constraint.

Strategies to address the risk of summer overheating

In an ultra-low-energy build it is unlikely that you will be cold in winter; the challenge is to ensure you do not overheat in summer. To meet the Passivhaus standard, you are limited to overheating (above 25°C internal temperature) for 10 per cent of hours in a year. This would actually represent an uncomfortable amount of time, so we would suggest you adopt an even more stringent target, perhaps 5 per cent maximum, and of course aim for as close to 0 per cent as practicable. The Passivhaus aim is to minimise the 'cooling' season, just as you aim to minimise the 'heating' season. In a cool-temperate climate such as in the UK, it is possible to eliminate the 'cooling' season altogether (in terms of the need for any mechanical cooling systems), and to manage the overheating risk through solar shading and your passive ventilation strategy (see Chapter 12). The design and layout of the windows is critical in addressing this successfully, as they determine how much summer solar gain you will have and should also contribute to your summer ventilation strategy.

Solar shading

The key to a successful design is to optimise winter solar gain while also avoiding too much solar gain in summer. The depth of the window within the wall (the reveal depth) will affect how much solar gain is achieved – the deeper your reveal, the less solar gain. If your window is set too deep in the wall, you may get very little winter solar gain – although you can consider splaying the side reveals. The position of the window in the wall is likely also to affect how optimised your installation psi-value is. (Ideally you should ensure that the wall insulation overlaps your window/door frames by 60-70mm. Usually, because windows tend to open inwards, this will have to occur on the outside of the frame.) These two factors need to be balanced against each other. If modelling in the PHPP, you can play with different depths to see the effect on the overall energy balance, which is very useful.

For those considering a Certified Passivhaus, the PHPP will adjust the effective shading of a window to reflect whether you choose window shading solutions that rely on manual handling or not – only a fixed or electronic shading device will be effective 100 per cent of the time. Even if you are not aiming for certification, it is best to assume that manually operated shading devices will not be 100-per-cent effective.

A variety of strategies for providing solar shading in summer are described overleaf – decide on which you prefer at an early stage and design solutions in! Large east- or west-facing windows are likely to be the worst culprits for overheating risk: this is because the sun is low in the sky morning and evening, and will therefore shine more directly through the windows. Large windows in small rooms will also increase the risk of overheating. It's important to be aware that internal blinds have a small value in reducing overheating (say, 10-20 per cent).

External window blinds

There are several options available for external blinds, which are all very standard in European countries but unfamiliar in the UK. The aesthetic is therefore a little alien and some solutions are likely to be problematic in conservation areas. Blinds can be manual or electrically operated and/or linked to a sensor. If linked to a sensor the blind may go up and down many times during a cloudy/sunny/windy day! The basic options are as follows.

- A separate external box, with side runners/guides for the blind, which is fixed on to the finished wall above the window. The larger the window, the larger the box to contain it. If externally insulating, consider how the weight might be supported without creating a point thermal bridge.
- A separate external box that can be butted up to the outside of the window frame and built into the wall – the blind can then run within the window reveal, which is a neater solution. Some window manufacturers offer particular models that integrate well with their window frames.
- A double-glazed unit with an external third pane of glass, which conceals an integral blind (pictured opposite). The U-value of such a window will be compromised but the aesthetic result is extremely neat and may be worth considering in some locations, where the overheating risk must be addressed but aesthetics are also essential. We have used

Internal and external views of windows at Denby Dale. Large glazing areas must be managed carefully for overheating – here large external electronic blinds are used. Images: Green Building Store

Section through double-glazed window with third outer pane (Internorm®), allowing external shading integral blind to be located in the gap.

these in the Totnes Passivhaus and they do look very smart and discreet. However, the gap for the blind can accumulate flies and increases cleaning!

The external blind boxes that are built into the wall should come with integral insulation to minimise thermal bridging. This will only mitigate the effect and the box will almost certainly increase your installation psi-value (i.e. worsen it). With a box fixed on the outside of the wall, there should be no thermal bridge issues, only visual.

Overhanging roof eaves

It is possible to design extra-deep eaves that will allow in low winter sun but cut out high summer sun. Having extra-deep eaves does create a particular 'look', again familiar elsewhere in Europe but not common in the UK. This is a solution for south-facing windows, not those facing east or west. Planners will need to consider this as part of a necessary design shift as we move to more energy-efficient buildings.

Independent canopy or veranda

A ground-level canopy or veranda to around 1m depth could provide good summer shading, although this should be optimised for your exact orientation and climate. The effectiveness of any canopy will be determined by its height above the window, as well as by the depth of the window reveal in your design. A framed veranda with green vegetation is another solution if the vegetation is verdant in summer only. Any canopy/veranda should ideally be structurally independent from the main house in order to avoid thermal bridging of the insulation layer.

Trees and other structures

Trees in the right location, especially those shedding their leaves in winter, could be a good solution. A wall or adjacent building could also provide some shade. Approach shading solutions creatively: using planting can be quite effective (and inexpensive).

Choosing the best strategy

Our advice would be to design in fixed shading solutions, such as overhanging eaves and canopies – these can have an aesthetic value as well as serving a useful function. Most importantly, remove overheating risks by placing glazing intelligently and utilise the power of the PHPP to test this, otherwise solutions could look tacked on or require overly complex detailing.

The PHPP includes a Shading-S worksheet (see Chapter 7, page 107), which requires data input for each window orientation and any shading (trees/buildings, etc.; window reveals; canopies/eaves). There is an average shading reduction factor of 75 per cent, which is the PHPP default value before you fill in the sheet, but this may be significantly different from your actual shading factor. Don't rely on this default to predict your solar heat gains. The rigours of the window and shading worksheets in PHPP are a necessary, if perhaps tiresome, part of the design process, as overheating is a genuine risk. In fact these sections of the PHPP are some of the most input-intense, which reflects their relative importance.

Summer ventilation

As discussed in Chapter 12, a Passivhaus adopts a 'mixed-mode' ventilation strategy. This means you are normally adopting passive strategies for summer cooling using windows and any roof lights. Simply using tilted windows on opposite sides of a building is the simplest method, usually with airflow from low to high level. When designing for effective summer ventilation, also make sure your solution is rain-, child- and burglar-proof!

There is a selection for your night ventilation strategy ('flushing') in the Summer worksheet in the PHPP. The mechanical ventilation with heat recovery (MVHR) would be set to '**summer bypass**' mode or switched off in summer, which means the heat exchanger (or heat recovery function) is not being used and fresh air is delivered direct to your rooms instead. This would normally be the case when external temperatures range from 15°C to 25°C. It should be noted that the summer bypass mode is not a cooling system and should not be relied on to act as such. In Germany the general approach has been that for any significant number of hours where the temperature rises above 25°C, you would start to use the heat exchanger again (this will minimise the air temperature being brought in) and then keep the windows closed during the day and open them at night for cooling.

Alternatively, you could install a **ground source heat exchanger (GSHX)**, which will pre-cool the incoming air. A typical system (brine-to-air) uses pipes filled with brine that are laid in the ground, where at depths of more than 1m the temperature remains static through the year – this 'coolth', or coolness, is then transferred from the brine to the incoming air prior to it entering the air handling unit of the MVHR system. Air-to-air GSHX systems also exist, but are now becoming less common in Europe, as condensation can form in the underground pipes, and if condensate cannot drain away adequately there are potential health issues. If you do decide to install a GSHX system, we would suggest not using an air-to-air system for this reason. GSHX systems were often installed in early Passivhaus homes, but they are far less common now for milder climates, where the benefits-versus-costs balance is less clear (although if you can install the pipework yourself, the economics change). A brine-to-air heat exchanger also works effectively as frost protection in winter (see Chapter 12, page 198). If the climate is more extreme (hot or cold), then the use of a system such as a GSHX becomes increasingly beneficial to avoid the use of high-power cooling devices, which would undermine the Passivhaus low-energy approach. In a maritime climate (i.e. less extreme) such as in the UK, sensible window design and appropriate shading should eliminate the need for any mechanical cooling.

In a more complex project – say, a block of flats – it may be necessary to look separately at some of the flat units, as overheating risks could vary significantly from flat to flat. This might require more expert input, not just the PHPP modelling.

The thermal mass of the building

The risk of overheating is influenced, to some degree, by whether the build is lightweight (timber) or of more massive construction (brick/block/concrete). The thermal mass of a body of material refers to its ability to absorb, store and subsequently release heat (due to its specific heat capacity and its mass); heavyweight construction materials, such as brick and stone, have a high thermal mass, while lightweight materials such as timber do not. Thermal mass is useful when the heat transfer between the material and the interior air roughly matches the daily (24-hour) heating and cooling cycle of the building (often referred to as the 'diurnal temperature cycle'). The material will then absorb heat during the day and release heat during the night, and this will effectively dampen the internal temperature variation, helping to stabilise it. Some materials (e.g. steel) can absorb heat very effectively but will release it too quickly, so you need materials with moderate thermal conductivity but a high specific heat capacity. Data needs to be input into the PHPP Summer worksheet to reflect this (see Chapter 7, page 107).

You don't need huge thick masonry walls to achieve effective thermal mass; something like a 25mm clay plaster can be sufficient to help dampen temperature swings between day and night. We would suggest that some provision is made for thermal mass in the design; the ground-floor construction can be a good option. Larger buildings will generally have a greater need for thermal mass than single houses.

Ordering windows

Finding someone well informed who can talk you through the various options is not easy at present, unless you can speak German fluently! It is worth spending time selecting your preferred manufacturer early on in the design stage and then optimising the specifics of the window (glass type, width, g-value, etc.), as this can save a lot of money. The PHPP will help with this, since you can see the immediate effect of tweaking elements of your design – 20mm extra on your external insulation, for example, will

probably be much cheaper than cranking up your window specification. Remember that order periods can be quite long for doors and windows – typically 8-12 weeks.

There are UK and Irish companies beginning to supply Passivhaus-suitable windows and doors, although generally these have been sourced from the Continent (one or two are assembling in Wales and Ireland). It is hoped that products will soon also be designed and made in the UK.

The construction phase

Installing windows and handling such expensive items on-site does demand extra care and attention. Triple-glazed windows (and insulated doors) can be quite heavy and, being costly, need careful protection so may need lifting equipment to install. Some are not necessarily individually pre-wrapped for delivery to site and you may want to stipulate this requirement when you place your order. As you will already have appreciated, fitting is not the same as a standard window installation and this needs to be properly understood by the main contractor, so that the installation sequencing runs in a practical and efficient manner. The windows and doors are both critical for the final thermal performance and the airtightness of the design; therefore the window subcontract becomes a package of work that needs early consideration at the planning and construction phases.

Window subcontractors

Think carefully about your strategy for window installation and seriously consider retaining interest in selection of subcontractors if they are to be used rather than the main contractor. There will be very few UK window installers that have any experience in installing for a Passivhaus build; even those recommended by your window supplier may not be that experienced! Some manufacturers will be able to supply and fit or may offer an advisory service, which would involve a qualified technician attending your site to demonstrate how a window should be fitted – a minimal one-day service. This is a very effective method of training your fitters (who may be more local to you); with reasonable competence they can then replicate the demonstration when fitting the remaining windows/doors. The alternative is to negotiate that one or two of your selected installers attend a training course (preferably on offer from the manufacturer). These courses are much needed and well worth an installer's investment. They could vary from two to three days to two weeks, depending on the manufacturer. Some also include training on repairs and maintenance, which might well be the best option on multi-unit schemes where a longer-term interest is to be retained, e.g. registered social landlords (RSLs); those trained can then train others on-site and ensure that quality is maintained.

It will take time for an established network of trained installers to develop across the UK, but in the meantime do not let your window installation just 'happen' – retain control.

Future developments in windows

Low-energy windows have to date had rather chunky frames, due to the additional insulation and airtight seals. Newer windows are already slimmer, and this trend is bound to continue as the technology develops and becomes more familiar. As noted at the beginning of this chapter, we would also expect progress in the development of outward-opening designs.

It is also true that low-energy windows generally do not meet more historic aesthetic requirements, especially as larger single panes are always going to be more energy efficient. Small window panes, with numerous window bars or mullions, are the common aesthetic in many conservation settings, and, while they generally perform poorly, we do seem to like the effect and our nostalgia proves to be rather strong! One or two window manufacturers are beginning to respond to this by designing windows that are aesthetically suitable for many conservation settings while still delivering the level of energy performance needed in an ultra-low-energy building (see bottom photo on page 77). Whether culturally we need to 'move on' is worth asking (the original reasons we had small panes of glass are long gone), but the reality is that there will always be a demand for replicating the historic look. Of course it can be only a 'dummy' effect, but even this will affect window performance to some degree – in particular reducing solar gain. If you want to go down this avenue, you will need to carefully assess the effect of compromised performance.

Window costs

At present, high-performance windows are expensive, but costs are already reducing. Such windows are essential and integral to the Passivhaus or an ultra-low-energy approach, in terms of both energy and comfort/health. The quality of these windows is extremely high and they must be seen as a long-term investment – in the UK, windows are generally replaced more frequently than is desirable, but a Passivhaus window will last a lifetime. See Chapter 2 for a discussion of Passivhaus economics.

Chapter Eleven • Windows

Conservation triple-glazed 'sash' window at Princedale Road, London.
Images: Princedale EcoHaus Ltd

RECAP

Passivhaus-suitable windows and doors have much better thermal performance than typical new double-glazed and poorer-performing triple-glazed units. They also are relatively leak-free (from air) when closed, if installed correctly. The modelling of a Passivhaus includes a thorough consideration of the whole-window U-value (the glass, the frame and the glazing spacers) as well as the effect on performance of the window installation.

Windows also deliver important 'free' solar energy into the house during winter. In summer, overheating must be controlled by appropriate devices. The shading requirements can be determined by the PHPP. A well-thought-out shading strategy is vital for any ultra-low-energy house.

It is important to ensure that you have designed in a natural cross-ventilation path for summer conditions. At night especially, you are sometimes likely to need to naturally ventilate and cool your home. An openable window at a high level in your staircase can be a good option.

Make sure you get your windows ordered in plenty of time and plan for them to be fitted by someone who has some experience of an ultra-low-energy build. Some suppliers offer an initial 'training' installation day if your contractor is unfamiliar with the process. These windows and doors represent a large investment, so ensure a good design detail (for minimum thermal bridging) and proper installation, or they will not perform to their potential.

CHAPTER TWELVE
Ventilation

Ventilation in UK housing and in Passivhaus, indoor air quality (IAQ), humidity, mechanical ventilation with heat recovery (MVHR) (components, heating, efficiency, noise levels, possible objections to MVHR, installation skills)

Every new or retrofit building requires a considered ventilation strategy. The basic principle is to ensure that indoor air is being refreshed regularly, although the benefits of achieving a good ventilation strategy are multiple – from avoiding mould growth to limiting airborne pollutants. The quality of the air will also affect how comfortable we feel – too hot, too cold, too dry, etc. Some effects of a poor ventilation strategy, such as cold draughts, overly dry air or noisy extraction fans, will always be noticeable. But it is also important to be aware that we are not always able to assess air quality with our own senses. Bad odours can be very unpleasant, but are often harmless, while more serious health issues relating to ventilation could well go unnoticed.

If you are commissioning or are part of a team delivering an ultra-low-energy building, you must be confident that your ventilation strategy will be effective. The Passivhaus standard adopts a 'mixed-mode' ventilation strategy, comprising a mechanical ventilation system, which includes heat recovery from the exhaust air (energy that would otherwise be wasted to the outside), and summer ventilation/cooling using openable windows. This combined approach is an integral and essential part of Passivhaus low-energy design and is considered part of the quality standard of a Passivhaus building. It has been adopted not only to address functional needs but is also part of the comfort standard.

Our experience is that a Passivhaus home will effectively extend the seasons in which you feel thermally comfortable with open windows – so you are likely to still be opening your windows after your non-Passivhaus neighbour has shut theirs. Because a Passivhaus is so airtight, even with open windows it retains its heat for longer than a conventional house, and in the shoulder seasons it will be getting significant solar gain.

Ventilation strategies in UK housing

Historically, buildings have been so 'leaky' that there is more than sufficient ventilation for occupants by default. Resultant draughts have been exacerbated by the common use of open fires, which further draw in cold external air in order to feed the combustion process. The traditional design of the high-backed 'wing' chair was adopted to provide shelter from this draught while seated facing the open fire. Up until the 1970s, people were content to live at much lower temperatures and would dress accordingly: in winter you simply wore lots of layers.

Typical housing in the UK from the 1920s onwards would be built using a cavity wall construction, consisting of two separate leaves of brickwork (later the inner leaf was changed to concrete blocks). About 80 per cent of existing housing is estimated to be of cavity construction. Cavity insulation started from around the 1970s, and, with the introduction of insulation standards, recent new housing stock does provide some improved thermal performance. However, this is still far below what a low-carbon economy would require. Only in 2006 was a statutory value set for maximum air leakage. This figure – a maximum air permeability of $10m^3/hr/m^2$ at 50Pa (see Chapter 9) – is not demanding and we doubt many new houses would fail this requirement. Currently, the typical airtightness level achieved is an air permeability of around $7m^3/hr/m^2$ at 50Pa. Conventional methods for ensuring adequate ventilation in the UK generally take the form of trickle ventilators in windows (often left in the closed position because people are not aware of what they are) combined with localised mechanical extraction from bathrooms (usually on timers, often noisy and sometimes disabled by the disgruntled) and the kitchen (generally via a kitchen cooker hood extracting to the outside). Such mechanical extraction clearly helps to remove smells and moisture at source, and to reduce condensation risks. However, it does not generally involve any heat recovery from the extracted warm, moist air.

Low-energy design necessarily moves us towards designing less leaky buildings. This minimises energy loss to the outside, improves insulation performance and reduces moisture ingress into the building fabric. The current airtightness target in the UK for a zero-carbon house is a maximum air permeability of $3m^3/hr/m^2$ at 50Pa – a target that still allows the use of natural ventilation strategies, i.e. openable vents or windows (without any mechanical assistance) and reliance on naturally occurring pressure differences to move the air through the internal rooms. Often the air will be vented out at a high level, since warmer air naturally rises, so the staler air might perhaps exit at the top of a staircase, for example. Of course, adopting natural ventilation solutions carries with it an energy penalty, and this has to be reflected in the energy targets set for space heating for such houses. The zero carbon standard requires $39-46kWh/m^2$ per year for space heating (depending on house type) – more than treble the Passivhaus standard (see Chapter 5, page 67). However, this would still be extremely low compared with the average performance achieved by current new UK housing.

Passivhaus ventilation – a mixed-mode strategy

The Passivhaus 'mixed-mode' ventilation strategy ensures that there is always a continuous supply of fresh air even when windows are closed. This enables the Passivhaus air leakage target of 0.6ach (air changes per hour) at 50Pa – the lowest of any building standard in the world. A common misconception about Passivhaus is that windows are physically non-openable or that you are not allowed to open them! This is not the case, and in summer the whole house can be ventilated naturally, just like any other house. In fact, a Passivhaus should be designed so that air can naturally flow through the building. As we saw in Chapter 11, openable windows can also be used effectively at night to cool down the building in the hotter summer months; this, combined with high insulation levels and appropriate window shading, is a core strategy for keeping low-energy houses cool. In fact, the PHI requires at least one openable window per room.

The mechanical ventilation with heat recovery system, or MVHR, is also sometimes referred to as comfort ventilation or heat recovery ventilation (HRV). MVHR units are not heat pumps and contain no refrigerants, which is another common misconception. MVHR systems simply handle incoming and outgoing air.

MVHR is a 'whole-house' ventilation system, with fresh air delivered to all the habitable spaces (bedrooms and living areas) and extracted from the wetter/smellier rooms in the house (bathrooms, toilets and kitchens). The air circulates naturally and gently from these supply zones to the extract zones, so that the whole house is continually being refreshed with clean, filtered outside air.

The heat recovery element of the ventilation strategy is key to Passivhaus, in that the incoming air is preheated by the extracted air. The extracted air is not mixed with the incoming air; the two air flows are separated by a thin plate with a large surface area through which the heat is transferred, in what is called a counter-flow heat exchange chamber (see Figure 12.1 opposite). In a Passivhaus the MVHR system is designed to ensure that the minimum temperature of the air delivered into any room is 16.5°C when the outside air temperature is -10°C.

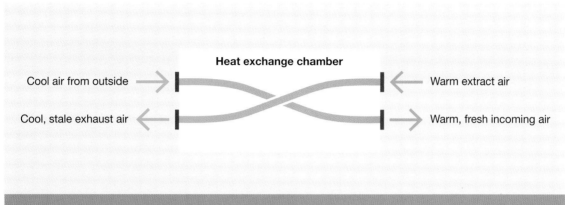

Figure 12.1 Recovery of heat in an MVHR heat exchange chamber.

In addition to 'recapturing' energy that would otherwise be wasted, this pre-warming of the supply air provides a level of user comfort, since it eliminates cold draughts. Cold draughts make us feel less comfortable (the wind-chill factor) and we tend to want to increase the ambient room temperature to compensate. With pre-warming of the incoming air, an average occupant will feel as comfortable with the room temperature set two degrees lower than in a conventional building.

Ventilation and indoor air quality (IAQ)

There are relatively few academic studies on air quality within the home, which is surprising given that such a high percentage of an average person's time is spent indoors. A typical person in industrialised countries such as the UK and US now spends 90 per cent of their time indoors. In these societies there is also a worrying increase in the number of people suffering from asthma – the UK and US have the highest levels of sufferers in the world. The possible connection between this and the rising use of chemical products in the materials, finishes and furnishings of our homes and buildings is commonly alluded to (as is a possible link between asthma and outdoor pollutants, e.g. from traffic). If there are grounds for making this connection with indoor allergens – and many argue that there are – then the introduction of continuous and controlled air supply must be of benefit. If this approach is combined with the use of natural materials and finishes and/or materials that emit low levels of volatile organic compounds (VOCs), such as concrete, linoleum, wood, ceramics, lime, clay and natural paints, then air quality must be improved. Using more natural cleaning products will also contribute to lowering VOC levels. Even without an absolute proven connection with health conditions, organisations such as the US Environmental Protection Agency (EPA) recognise that poor IAQ is a prominent environmental problem, with a related health penalty for society to pay. The US LEED standard dedicates a whole section to IAQ, mostly limited to discussion of materials and finishes, with minimal reference to ventilation strategies. In order to ensure good IAQ, both these aspects need to be considered.

Since buildings will become increasingly airtight in the future, the risk of poorer IAQ must be a serious consideration. Some people have been interested in using low-embodied-energy, locally sourced and natural materials, and have built eco-houses without necessarily attempting to build leak-free constructions. This approach will certainly have improved air quality without the use of typical ventilation strategies (i.e. trickle vents in windows and bathroom extraction fans). However, if your intention is to build a low-energy (in use) house, then the ventilation strategy becomes critical.

Passivhaus ventilation and indoor air quality (IAQ)

Measuring carbon dioxide (CO_2) levels in the indoor air is considered an acceptable indicator of air quality, since CO_2 levels reflect indoor air pollutant levels. Air quality is classified according to the European Standard EN 13779, with four levels set at different CO_2 concentrations, from IDA 1 (high quality) to IDA 4 (low quality). IDA 2 (medium quality – 400-600ppm [**parts per million**] of CO_2) is considered to be good air quality. Both ASHRAE (an international research/ standards organisation dealing with ventilation) and the US Occupational Safety and Health Administration (OSHA) set a maximum indoor CO_2 level of 1000ppm. This compares with typical outdoor levels of 350-450ppm. People start to complain about stuffiness and odours as levels rise above 600ppm (IDA 3 is 600-1000ppm – moderate quality), and drowsiness kicks in as they rise above 1000ppm (IDA 4). There is a useful study of carbon dioxide levels relating to a Passivhaus retrofit of a block of flats in the Marbachshöhe area of Kassel in Germany.[1] It compares winter CO_2 levels in a standard bedroom in the original building, which relied on natural ventilation, with a bedroom in the Passivhaus retrofit, which relied on MVHR. The resultant measurements of CO_2 (see Figures 12.2 and 12.3 opposite) show the levels regularly rising above 1500ppm in the old bedroom while remaining comfortably below 1000ppm in the Passivhaus, mainly within level IDA 2. Figure 12.2 clearly shows that opening windows will quickly purge any stale air, but once they are closed the CO_2 levels rise again quite rapidly. The bedroom in the old building has particularly poor air quality during the night. A second study compared Austrian low-energy housing, relying on natural ventilation, with comparable Passivhaus units.[2] Here both had acceptable air-quality results (although the Passivhaus was more energy efficient), and the MVHR-ventilated units had lower CO_2 concentrations, in particular during cold periods and especially in multiple-occupancy apartments.

The Passivhaus approach is often applied to school buildings in parts of Europe, e.g. all new school buildings in Frankfurt must now be built to the standard. The first Passivhaus school in the UK is now open in Exeter. Part of the motivation for adopting the Passivhaus standard for schools is air quality, which is known to improve concentration and learning capacity.

The PHI has calculated that the IDA 2 standard will be met by an airflow rate of approximately 30m³/hr per person – the level recommended by the PHI, and the default assumed in the PHPP.

Automating the MVHR – CO_2 sensors

In a house with large daily or weekly variations in the number of occupants, it is possible to link the MVHR unit to a CO_2 sensor that will vary the air change rate the MVHR unit delivers according to need. This is a standard approach in non-domestic buildings and could also be used in tenanted residential buildings.

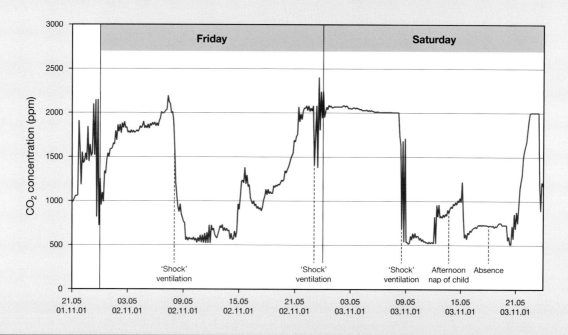

Figure 12.2 Levels of CO_2 in an apartment using opening windows as the ventilation strategy.
Adapted from: Passivhaus Institut (PHI) 2001/2. CEPHEUS: 'Measurements and evaluation of Passive House apartment buildings in the Marbachshöhe neighbourhood of Kassel, Germany'

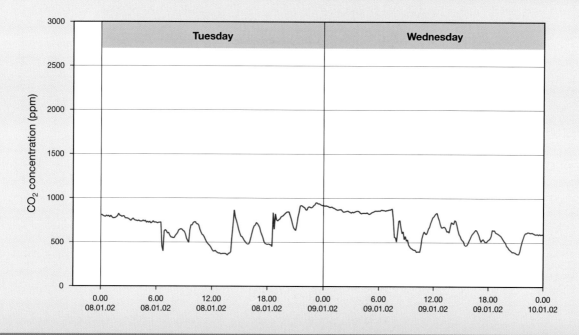

Figure 12.3 Levels of CO_2 in an apartment using MVHR as the ventilation strategy.
Adapted from: Passivhaus Institut (PHI) 2001/2, as above

Montessori school in Aufkirchen, Germany, built to the Passivhaus standard. *Image: Passivhaus Institut*

There is also a PHI-recommended air change rate of 0.3ach at normal air pressure. This is set to avoid the risk of low air humidity in the winter (dry air), which reflects the climate conditions in Germany.

Indoor air quality (IAQ) and relative humidity (RH)

Relative humidity levels are also a factor in maintaining comfortable air conditions for occupants. Air needs to be neither uncomfortably humid nor uncomfortably dry. The RH target should be between 35 and 55 per cent: 30 per cent is considered very dry air, while 70-80 per cent is very humid – at both these levels there are implications for health (see Figure 10.2, Chapter 10, page 149).

The amount of water vapour that can be contained or carried in the air is relative to its temperature (hence the term 'relative humidity'). We are sensitive to humidity levels because our bodies use evaporative cooling to regulate our temperature. The rate of evaporation of our perspiration changes depending on the RH, so we may feel cooler or warmer at the same air temperature. In warm, humid weather, the rate of evaporation slows, so we will feel much hotter than in warm, arid conditions. Once air reaches 100-per-cent RH, the air is 'saturated' and condensation will occur.

Dry air

Drier air can be an irritant to a building's occupants, and tends to exacerbate respiratory diseases. If external fresh air contains too little moisture (say, on a very cold winter's day), it is useful to be able to modify the delivered internal air quality by setting the MVHR to a lower airflow rate (i.e. the PHI-recommended minimum of 0.3ach). If the external air is already dry, then the MVHR can exacerbate this (warming the air also dries it!). The Passivhaus standard originated in a climate that tends towards drier winter air conditions, and for this reason the PHI advises against the oversupply of fresh air, recommending that airflow should not be increased unnecessarily. In such conditions, drying clothes inside your building can be useful, to add moisture back into the air.

In very cold, dry climates it is also possible to install membranes that will allow the exchange of moisture from the extract air into the supply air. In the USA, ventilation in low-energy housing is delivered by entropy recovery ventilation (ERV), now often referred to as energy recovery ventilation, which includes this moisture-exchange function. These systems are used where there is either high humidity or very low humidity levels.

Humid air

If, on the other hand, the air becomes too humid, dust mites proliferate, the risk of mould growth increases, and bacteria and viruses can flourish. A maritime climate such as in the UK tends to higher humidity levels, and in these conditions it may be useful to set the MVHR to slightly higher airflow rates (for a target air change rate of perhaps 0.35-0.4ach).

There are also other useful mechanisms for modifying moisture levels. Indoor plants, for example, are great natural humidity modifiers. It is also worth thinking about the choice of surface materials in the house, and considering those that can readily absorb moisture from the air and then release it back in a cyclical process. Hygroscopic materials, such as natural clay and lime plasters, will hold and then release water vapour cyclically into the surrounding atmosphere, and can also permanently absorb some VOCs. This dynamic 'hygric' (moisture-related) response will keep the RH more stable, helping to avoid surface condensation or mould growth. The quality of the air experienced in rooms with clay-plastered walls is often commented on by those inhabiting them. There are academic studies that demonstrate the effectiveness of using vapour-open and hygroscopic materials,

which most definitely help to improve IAQ when used as part of a holistic approach to ventilation. We would encourage this approach. Remember that the initial moisture levels in a new build will be affected by the moisture within the structure itself, and it may take some time before the building fully dries out (a minimum of two seasons, i.e. years). Moisture levels will also be determined by the form of construction; some techniques involve more water than others.

Monitoring internal RH levels in Passivhaus homes built in a maritime climate (using humidity/temperature data loggers) would be a useful exercise, both in terms of building up a data resource and for the purposes of furthering optimisation of the overall ventilation strategy in any given building.

Humid air, mould growth and ventilation

Condensation forms on surfaces where the air is carrying high levels of water vapour (See Figure 10.1, Chapter 10, page 148). This, combined with low surface temperatures, may eventually lead to mould growth. Often this can occur where air is not being regularly refreshed (mould in room corners, behind furniture or in built-in cupboards is common). Moisture arises from human breathing/sweating, cooking, washing, plants, etc., and occupational behaviour can exacerbate potential mould problems – for example, keeping windows closed, not using bathroom extractor fans (in conventional houses), and drying wet clothes over radiators. With the trend towards higher internal temperatures in our homes, mould growth is not an uncommon problem. However, providing continuous background ventilation (both air supply and air extraction) alleviates most of these risks. The Passivhaus approach of continuous insulation, minimal air leakage and triple-glazed windows further ensures that the internal surface temperatures, even at window panes, do not fall below 17°C. This not only eliminates surface condensation but also contributes to avoiding convection-driven draughts.

On retrofit projects, be aware that once air leakage is radically reduced, air humidity will rise – with an increased risk of surface condensation unless a ventilation strategy is in place. Any area that is not insulated to a good level may then be vulnerable to mould growth. This risk is addressed as part of the EnerPHit standard (see Chapter 1, page 30).

Mechanical ventilation with heat recovery (MVHR)

A Passivhaus is designed so that the MVHR system can provide all the ventilation needs during the winter months, when most people generally do not want windows open. The MVHR system will usually achieve an air change rate of 0.3-0.4ach under normal conditions. As already noted, the PHPP software recommends an airflow rate provided by the ventilation system of 30m^3/hr per person as a design standard. The design default for a typical house assumes 35m^2 of treated floor area (TFA) per person. This may not reflect a larger house or very small flat, so the PHPP should be modified to reflect these circumstances (see Chapter 7, page 93).

The MVHR system is constantly delivering fresh air to a building's habitable spaces and extracting air from the wetter areas, all at relatively low air speeds. There are typical recommended extract rates – 60m^3/hr for the kitchen, 40m^3/hr for bathrooms and 20m^3/hr for showers and WCs (see Figure 7.10, Chapter 7, page 105) – which are similar to those in current (2010) UK Building Regulations (England and Wales). There are also typical recommended supply airflow rates for the different rooms (living room 40m^3/hr; master bedroom 30m^3/hr; other bedrooms 20m^3/hr).

These are not exact requirements and you can adjust these to reflect room size and likely occupancy. The extract and supply air volumes need to be in balance with each other for the system to run efficiently (total supplied airflow = total extracted airflow), and the PHPP Final Protocol supplementary spreadsheet helps to ensure that this is the case. Balancing the system prevents air being pushed through the building fabric rather than through the MVHR system. To allow the movement of air through the house, '**transfer paths**' are required. These are most commonly and simply achieved under doors, with a larger-than-normal (20mm) gap. Alternatively, a more discreet option is to adapt the architrave detail at the door head (see Figure 12.4 and photo opposite). If a planned transfer path was not included, the air supply would not be dangerously affected (doors are opened fairly frequently in any case); this is just not ideal.

There will also be intermediate spaces or 'transfer zones' – corridors and stairwells – where there is no direct extract or supply (see Figure 12.5 opposite).

The rooms you will be supplying air to tend to be the larger ones in any home (i.e. bedrooms and living spaces). This means that the air is generally flowing from larger to smaller spaces (e.g. bathrooms) and you therefore tend to achieve higher air change rates in the extract rooms, where this is needed to remove moisture and smells – rather convenient!

Incorporate the MVHR system into your scheme at an early design stage in order to ensure it works efficiently and effectively – try to minimise duct runs and bends, for example. The central unit itself can be located in a utility space, loft space, external store/garage or in a cupboard. The typical unit size for a small- to medium-sized house (60-120m^2 floor area) might typically be 800mm x 600mm x 400mm,

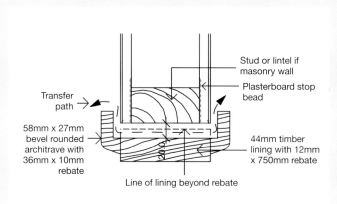 

Figure 12.4 Architrave section to create hidden 20mm air transfer path, as pictured here.

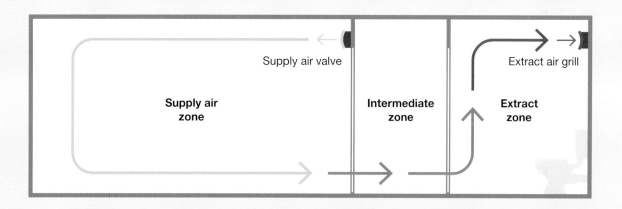

Figure 12.5 Supply, intermediate and extract zones for an MVHR system.

plus allowance for the connections of the associated ductwork. If the manufacturer can confirm that the central unit is suitable for an external location (outside the building's thermal envelope), rather than indoors, this can have some advantages:

- It can help to maximise the treated floor area (a services cupboard is not included in the TFA calculation in Passivhaus).
- It is more accessible if being maintained by a third party (e.g. in social housing).
- Any potential noise from the unit is outside the building (although this is very minimal).

However, an externally located MVHR unit would need suitable insulation and the condensate drain to be protected from frost, and it may not always be as efficient as an internally located unit.

MVHR components

The main components of an MVHR system are as follows.

- **Central unit** Contains the air-to-air heat exchanger and the two fans needed to push the air through the system, the two main air filters for the extract air (from the house) and fresh air (from outside). The unit should be heavily insulated, and will arrive pre-insulated if it is a Certified Passivhaus unit.

- **Condensate drain** Condensate is created in the heat exchanger as the extracted warm, moist air is passed by the incoming cold, fresh air: when the heat is taken out of the old warm air, it is no longer able to hold as much moisture and the water vapour it contains returns to its liquid state. The condensate drain needs be connected into the building's drainage system and will need insulating if outside. Make sure you have access to inspect it if necessary. Apparently you can also obtain a condensate evaporator, if access to a drain is impossible.

- **Frost protection** A frost-protection strategy is needed alongside the heat recovery system, regardless of whether it is located internally or externally. This can take different forms, but the most common approach is a small electric element located on the intake duct (if the unit is indoors, as is most common, this will be between the MVHR unit and thermal envelope – see Figure 12.6 on page 200). The air entering the MVHR central unit must never drop much below zero. Ensure that the separate thermostatic sensor is located in the intake duct between the frost protection element and the MVHR unit. (A common mistake is to place the sensor between the element and the thermal envelope.) The element will preheat the fresh outside air if there is any danger of this cold air freezing the condensate that forms in the heat exchanger. If the condensate were to freeze, the airflow would be at least partly obstructed and the system would not be able to function properly. Frost protection usually kicks in at -5°C, preheating the air to just below freezing. While still essential in the mild climate of the UK, it will operate only rarely. If you have a ground source heat exchanger (GSHX), this would deliver your frost protection.

- **Air filters** The central unit incorporates two paper filters – an F7 (fine-particle filter) for the outside incoming air and a G4 (coarse-dust filter) for extract air returning to the unit. These filter categories are common across all MVHR units. They serve to protect the unit but also ensure that the supply air is

Central unit with cover off and illustration of how cold and warm air is passing through the heat exchanger. (From top left to bottom left: Supply / Extract / Exhaust / Intake.)

clean and your house air relatively free from dust and pollen. This is another Passivhaus comfort benefit and excellent news for hayfever sufferers (around 20 per cent of the UK population) and asthmatics! Changing these filters is extremely simple, and you may need to renew them every three to six months, depending on your location (city or country). To check, just inspect the filter by eye – you will see when it's dirty. Make sure you switch off the machine while changing the filters, to avoid air contaminants entering the unit. If you are in an environment with relatively clean air, two G4 filters may be perfectly adequate, rather than one F7 and one G4, which would be a cost saving; but if you are thinking of this, discuss it with your MVHR supplier first.

There is a further filter for grease, which should be located at the kitchen exhaust outlet to stop grease entering your duct system. It should be removed and washed periodically to ensure it doesn't get so clogged that it affects the fan's ability to pull air through it. (This could be necessary every fortnight if you do a lot of frying!) Alternatively, you can get disposable fleece-based filters. Cleaning this filter is a particularly important maintenance task (unless the unit is designed to automatically rebalance between supply and extract), as a clogged filter will reduce the rate of extraction.

In a Passivhaus you would not use a conventional extractor fan/hood over your cooker, which extracts directly to the outside. Instead you would use a recirculating hood with removable carbon or washable filters. This will also help to remove grease and smells at source. Some people choose not to use a cooker hood at all, relying on the MVHR kitchen extractor alone. Whichever your preference, we would advise against locating

Dirty and clean air filters!

the MVHR kitchen extractor directly above the cooker.

There may also be a filter for the frost protector, which should be changed when the main unit filters are changed.

- **Fresh air intake (or 'ambient') and exhaust ducts** The location of these two ducts, which go direct to the outside, is an important consideration in terms of both aesthetics and performance. It is ideal if the two ducts can be oriented in the same direction, in order to avoid differential wind pressures. Do not locate the fresh air intake too near the ground; generally the air will be cleaner at height (preferably above 3m). In addition to a grille on the end, the duct should ideally be protected from rain and snow (the roof eaves can be ideal for this, or perhaps fit a purpose-made small hood). Best practice is to have a wider grille diameter than duct – use a duct reducer between the two. This will avoid any undue resistance or pressure losses in the airflow, which would then increase the power consumption of the fans. Finally, do not locate the two ducts adjacent to one another; an ideal spacing is a minimum of 2m. You don't want to create a short circuit between the two ducts! If placed one above the other, it is better to have the exhaust above the intake, at a minimum spacing of 1m. If you decide you need an insect filter fitted, make sure it is accessible for cleaning and replacement.

If the unit is inside the thermal envelope, then both these ducts will need to be insulated to a high standard along their full length (from the MVHR unit right up to the building's thermal envelope), to avoid condensation on the pipe (see photo, right). If uninsulated, these pipes also become two significant thermal bridges. There is a table in the PHPP where the depth of this insulation must be entered – normally 50mm. The lengths of these ducts should be kept to a minimum, and these values must also be inserted into the PHPP. The insulation material is a non-standard product that will not deteriorate in moist conditions (technically known as 'diffusion resistant'), e.g. Armaflex®. Make sure this type of product is what is used – it is expensive, so someone might try to skimp!

If not insulated, condensate will form on the duct surface. This picture was taken prior to insulation being installed.

If the unit is located outside the thermal envelope, then the above requirements for insulation apply to the supply and extract ducts (see below) between the unit and the thermal envelope – not the fresh air and exhaust ducts. (See Figure 12.6 below.)

- **Supply and extract ducts** These ducts run from your central MVHR unit to all the rooms of your house, often in ceiling and floor voids. You can use metal or plastic

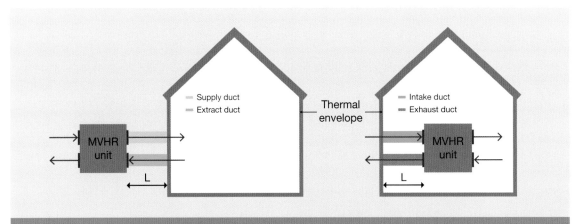

Figure 12.6 Insulation of ducts is dependent on the location of the central unit. Keep the length of these runs (L) as short as possible.

ducts in a variety of shapes – round, oval or square. Pipes should be smooth internally, as this reduces airflow resistance, minimising energy use. Keep duct lengths short and runs simple – not always easy with retrofit projects. Make sure the ducts are clean before (as well as during) installation. Open ducts on a building site will be filthy! Duct runs need to be airtight: using duct systems with proprietary seals makes the process easier and more reliable than sealing each connection by hand, sometimes in awkward locations. If not airtight, ducts will become dirty and the system will be inefficient. Some systems use single-length semi-flexible ducts (running separate smooth-walled ducts to each outlet location), which eliminate this risk.

- **Silencers (sound attenuators)** There should be silencers located between the central unit and both the first supply and first extract terminal, or outlet (see right). You should also provide a 'crosstalk' silencer between adjacent rooms on a duct run (since otherwise the ducts will act as a sound transmission service!), especially where privacy is most important. An MVHR system with separate semi-flexible ducts running to each outlet/extract location will serve this purpose.

- **Supply and extract air terminals** Terminals are fitted in each room to either extract or supply the air. Supply outlets can be fitted with jet nozzles, which will throw the air across the ceiling (if fitted approximately 150-200mm below the ceiling, measured to the centre of the jet). These are useful if the supply outlet must be located near a transfer path or an extract zone. Otherwise, both supply and extract outlets should be positioned as far away as possible from any transfer path, to optimise airflow. Outlet valves have different shapes, depending on which way the air is flowing – if the wrong valve is installed, it will result in a streaming noise. (This is an easy mistake to make.) It's also important to make sure the connection to the wall is airtight, so that air does not stream into construction cavities rather than the room! Avoid locating outlets near any obstructions. Outlets can be regulated to deliver or extract more or less air – normally this will be set at the commissioning stage and then fixed to avoid being easily tampered with.

- **Summer bypass** During the warmer summer months it is useful to bypass the central unit so that incoming air is delivered

Ceiling-mounted duct and silencer being 'boxed in', at the Camden Passivhaus.

Typical ceiling-mounted air terminal. The flow rate is adjusted by rotating the central disc.

direct to your rooms without heat recovery (although it is still filtered). This needs to be an automatic function to be useful. Without this feature it would only be practical to use the MVHR during the cooler months, i.e. you would need to switch it off completely in the warmer months.

- **Central control unit** This should be a simple unit – generally you will need only five basic settings:
  - standard setting – normal use
  - reduced operation (30 per cent less), e.g. when on holiday
  - increased operation (30 per cent more), e.g. when cooking or for parties!
  - summer mode (bypass or switched off)
  - off/unoccupied mode.

If the unit is complex, then it has been demonstrated that occupants are less likely to engage with the system. Some interaction with the system is needed, however, as it is important that users feel they have some control over it. Summer mode can mean either summer bypass (see above) or that the system is switched off completely. Different users will have different preferences.

A simple user control unit.

MVHR and heating

Passivhaus reduces the space heating requirements to such a degree that a conventional heating system is not needed. There is some residual need for additional heat during the coldest winter weeks, and one of the key early design decisions relates to the strategy for providing this. (See Chapter 1, page 24, and Appendix A for a discussion of space heating options.) One approach is to deliver heat using the MVHR system, using a small radiator (called a supply duct radiator or in-line duct radiator) inserted into the supply duct just after it leaves the MVHR unit, which adds a small amount of heat into the supplied air. In a Passivhaus it should be possible to meet the residual heating need in this way. For a typical house, the size of the radiator will be approximately $300mm^3$.

Some MVHR units will offer an integrated heater, located within the central unit. The heat source could be the hot water store, or it could be electrical or based on a heat pump or heat main. If this is your preferred space-heating method, then the supply ducts will also require insulation – normally 20mm of a standard quilted insulation such as glass wool or sheep's wool. There are also some suitable pre-insulated ducts on the market.

There is a possibility that the needs of heating might conflict with the air supply needs. The two requirements could be in conflict in a small room with high occupancy or a large room with low occupancy. But, given this caveat, using one system for the two purposes is an attractive solution and is fairly commonly adopted with success. Choosing to heat the air will reduce its relative humidity (RH), but this is of course true of heating generally.

If the exhaust air fan failed, cold outside air could flow right through to the heating coil and potentially freeze it. For this reason, if the fan

Supply duct radiator (with a mobile phone to indicate size).

MVHR compact unit at Princedale Road, with an extra hot-water-storage cylinder adjacent.

does fail, the whole ventilation will shut down automatically.

The supply air temperature from the radiator must not rise above 52°C, to prevent smouldering smells from any dust in the air within the duct.

Another option is an MVHR compact unit, in which the MVHR is usually combined with air-source heat-pump technology to supply heat (through the ventilation system) as well as to generate hot water. This approach was used at both Princedale Road and Lena Gardens (Victorian terraced houses in London). The system will normally have a hot water store, which could then be connected to solar panels. Some feel that this solution is overly complex, but it does save on space – hence the name – and provides a 'one-stop' solution.

Efficiency of the MVHR

There is a wide variety of companies that supply MVHR units, but for Passivhaus the efficiency of the unit is critical, and many currently on the UK market would not be Passivhaus-suitable. Some units have been certified by the PHI; others might be Passivhaus-suitable, but this is one component of a Passivhaus build where the PHI strongly advises selecting a certified unit. See Chapter 7, page 106, for the criteria for a Passivhaus-certified MVHR.

Efficiency of the central unit

The efficiency of the heat recovery must be above 75 per cent, but there are units that can achieve over 90 per cent efficiency. Take care when offered technical information by suppliers – the efficiency measurement for a Passivhaus-certified unit must be according to the testing method of the PHI and not that of other bodies or manufacturers, which are less rigorous. Along with the heat-recovery efficiency, the unit's electrical efficiency is key. This relates to the energy used by the fans to move a given volume of air. There can be significant differences between comparable units. The PHI stipulates a maximum electrical energy use of 0.45Wh/m³ (watt hours per cubic metre [of air moved]). However, the efficiency of the whole system will be strongly affected by the layout and design of the ducting and the commissioning process. Design and installation by ill-informed parties who are unfamiliar with the technology is likely to result in poor efficiencies that are well below the unit's capability. The efficiency as quoted assumes that:

- the whole system is balanced with ±10 per cent tolerance (supply and extract)
- all air goes through the exchanger and none is lost through leakage in the unit
- ducting layout is as per the design (as short as possible and sized correctly)
- airtightness in the building is as per the design intent
- there are no undue heat losses in the ducting due to poor installation/insulation.

MVHR efficiency and airtightness

All other factors being equal, an MVHR system will perform more efficiently the better the airtightness of the building. For an MVHR to be an efficient and appropriate choice, i.e. to improve energy consumption, it must be combined with a high standard of leak-free design. The general rule of thumb is that an MVHR should only be installed in buildings achieving an airtightness of 3ach at 50Pa or better. One recommendation[3] is that a maximum rate of 1.5ach at 50Pa should be the target to ensure efficient performance. This is also the statutory standard in Germany. Since the Passivhaus standard is 0.6ach at 50Pa, we would strongly advise that low-energy designers aim to achieve this target if MVHR is to be part of their selected ventilation strategy.

In a development of Passivhaus apartments in Hanover, Germany, it was calculated that for every 1kWh of electricity used by the MHVR unit, 16.5kWh of heat was being conserved through heat recovery. Even if you take into account the inefficiencies in electricity generation (primary energy), this still represents a very impressive 'coefficient of performance' (COP). A correctly installed unit in a typical house might use around 1kWh of electricity per day (equivalent to around 15p a day at current prices). If saving more than 16.5kWh of natural gas, for example, and assuming a gas price of 5p per kWh, clearly there would be a big net saving.

Obviously this efficiency does not apply if using an MVHR in 'summer bypass' mode, but a thrifty occupant can always choose to switch it off outside the cooler months and open their windows instead!

Efficiency and commissioning

The commissioning process for an MVHR system is very important and requires skill and knowledge. Make sure you use someone with a certain amount of experience to design the layout of the system and specify the components, as well as commission prior to handover. Generally, MVHR units have required a thorough balancing of the system at the commissioning stage, but some newer systems have self-balancing mechanisms, which simplify this process and ensure that ongoing balancing is maintained. Manual balancing will take some time – if experienced, perhaps half to one day – to ensure that everything is operating according to the design plan. If the system is not adjusted initially, it will not operate as you expect! Needless to say, the advantages of a whole-house mechanical ventilation system, such as conferring a more comfortable indoor climate, are only achieved if the system is designed, installed and commissioned to work as intended.

Part of the commissioning will be to ensure that the supply and extract air fans are properly balanced and that noise levels are within allowable limits. For this the intake and exhaust airflows will also need to be measured, as well as the wind velocity on the day of commissioning. On windy days, it cannot be accurately commissioned.

Each air inlet or outlet needs to be adjusted so that it is either extracting or supplying the amount of air intended (the maximum deviation is 10 per cent). The air valves in most systems will need adjusting several times, as changing one outlet affects all the others. Once complete, the outlet and inlet valves should be fixed using a locknut or similar so they cannot be tampered with. Using a 'flow-finder' device is generally accepted to be more accurate than other measurement methods, and is recommended by the PHI. This is based on zero pressure compensation, i.e. the airflow is hardly affected by placing the device in front of the outlet. A vane anemometer, on the other hand, works by being placed over the duct; the airflow stream then hits the vane, making it rotate. Some conversion of the reading may be necessary, depending on the units of measure the anemometer supports.

The PHI is encouraging the development of self-balancing systems, as these will avoid much of the complexity of commissioning described here.

Sound considerations

The PHI sets a maximum noise level of 25dB(A) ('A-weighted' decibels) in living spaces, so the system should effectively be silent at supply locations. At extract locations it should be virtually silent, at 30dB(A). The permitted noise level at the unit itself is slightly higher: 35dB(A) (still very low, but audible). Meeting these standards does require design input – it won't just happen.

Sources of sound could arise from 'crosstalk' between rooms (the sound being transmitted via the ducts) or from the unit fans or other structure-borne sound causing the ducts to vibrate. All of these can be designed out. Being aware of them should help anyone buying or specifying such a system to check that the installation has included the following:

- The supply ducts are isolated or 'decoupled' from the unit with a semi-flexible PVC pipe connector (not aluminium).
- A crosstalk silencer is fitted between every connected room, and between the unit and first supply and first extract outlet (this also helps with decoupling the supply ducts from the unit). Note that on some systems you would connect separate supply semi-flexible ducts from each extract and supply point back to a central acoustically lined plenum box, and therefore would not need silencers between rooms.
- The duct runs are kept as short and simple as possible (not always easy with retrofitting), which reduces the likelihood of vibration in the duct. This means minimising bends.

Some MVHR suppliers have told us that silencers are not needed, but we would not be content to install a system without such provision.

Fire considerations

If ductwork must cross through a fire compartment (e.g. between adjacent flats), then there are fire shutters available as well as cold smoke shutters, but in a single-family dwelling this should not be needed.

Objections to mechanical ventilation (MVHR)

In a conventional building, we already use mechanical energy in our ventilation strategies when we extract air from our bathrooms and kitchens. Using energy to draw fresh air into our homes, however, causes concern for some.

The energy use for the fans is regarded as wasteful in a low-energy building when a natural solution could be adopted. This argument is flawed, since, as we have seen, the energy saved by using such heat recovery systems far outstrips the small power required by the fans. Not only can heat be recovered from the exhaust air, but with mechanical ventilation the building itself can be constructed to be less leaky.

A further objection might be the energy used in manufacturing such a system. The ventilation system, as noted already, is integral to the overall low-energy strategy, which includes negating the need for a conventional heating system. Also, by reducing the overall energy requirements, any renewables (photovoltaic panels for electric energy and solar-heated water) used to achieve zero carbon can be kept to a minimum. In this way the energy invested in the system is more than compensated for by savings elsewhere.

Another, perhaps significant, anxiety is what happens if the system fails. The unit is a relatively simple piece of kit. If it did cease to operate, there would be no immediate repercussions, and windows can always be opened. Air is entering and leaving the building all the time (doors and windows opening and shutting, etc.). In summer the whole system can be turned off, in any case! A building is not a hermetically sealed bag, even with low air leakage levels. In winter, if the system did fail, you would eventually notice increased stuffiness and smells. The dangers relating to a faulty gas boiler are potentially far more serious, and most of us manage to have these in our homes without undue concern. MVHR units have been used widely on the Continent for many years.

Poor design / construction skills

Sadly, in the UK there are many accounts of MVHR systems being poorly installed. Common mistakes are:

- missing insulation on fresh air intake or exhaust ducts
- missing filters (especially the grease filter on kitchen exhaust)
- air intake adjacent to a pollutant source (e.g. compost heap or dustbin cupboard)
- poorly installed ducts (e.g. lots of twists and turns; ducts not clean when installed; not airtight)
- supply and exhaust airflows not balanced at commissioning (making the system inefficient)
- condensate drain poorly installed (no trap or an inadequate fall)
- supply and exhaust ductwork connected the wrong way round (or wrong outlet type fitted).

These are not complex problems and stem from general unfamiliarity with installing such systems. On smaller projects it is much easier to manage this type of quality issue; however, on larger commercial projects the risk of errors increases. If you are building in a culture where the levels of workmanship associated with achieving the Passivhaus airtightness standard are far from normal (certainly in the UK), or where MVHR is a relatively new concept, then there is a real risk of seeing badly installed and poorly performing systems, and this is likely to continue to be the case in the medium term. This will be exacerbated if training does not form part of the drive to achieve good-quality low-energy housing. As discussed in Chapter 4, in the UK we are extremely poor at providing contractor training. The government is moving swiftly towards legislating a zero carbon standard (due for implementation in 2016) without ensuring integrated training. As we will almost certainly see the increasing use of MVHR systems (perhaps becoming the dominant form of ventilation post-2016), training in specification and installation of these systems is something that needs to be addressed at national policy level.

As noted, the correct commissioning of the MVHR unit is crucial, and it would be sensible for the client / architect / Passivhaus Designer / Consultant to attend this. The advice in this chapter should help to ensure the basic mistakes outlined above are avoided!

Briefing the user

Part of ensuring a successful installation involves communicating with the occupant(s) so that they understand the basic principles and are clear about the requirement to change and/or clean filters. Modern units will alert you when to make filter changes. Everyone needs to understand what the ventilation outlets are, and that it is important they are not tampered with or blocked.

MVHR units should come with user guides, but far more important is that the user controls are intuitive, and at present this is not always the case. Given the simple nature of an MVHR unit, straightforward user controls should be achievable for manufacturers.

MVHR and the future

It needs to be appreciated that these systems are being continually improved and refined. It is likely that future units will be designed for a more straightforward commissioning process and it is hoped that certain components (particularly the frost protector and supply heater) might be integrated into the central MVHR unit – this appears to be the current trend.

RECAP

The use of mechanical ventilation with heat recovery (MVHR) is an essential part of the Passivhaus mixed-mode ventilation approach, and the PHI strongly recommends that the MVHR unit is Passivhaus-certified. There is no other viable ventilation solution at the low air leakage levels needed in an ultra-low-energy house.

The system is straightforward, but does require early design consideration and needs to be installed and commissioned properly. This is no more complex than correctly installing a gas boiler, but in the UK we do not yet have the equivalent wide knowledge base and experience.

With growing concerns about air quality and health, ensuring good indoor air quality (IAQ) is critical. Available research already indicates that a correctly installed and operating MVHR has a positive effect on IAQ and humidity levels. Passivhaus uses the European air quality level IDA 2 (CO_2 levels of 400-600ppm) to inform its MVHR operating parameters.

MVHR is essential only in the winter months, while in the warmer season you are likely to naturally ventilate your home by opening windows. It is usually possible to provide your residual space heating requirements through the ventilation system, using a supply duct radiator.

CHAPTER THIRTEEN
Living in a Passivhaus

Noise, energy bills, kitchen and bathrooom, drying clothes, the MVHR, entering and leaving the house, case studies (Totnes Passivhaus, Denby Dale, Grove Cottage, Passivhaus apartment buildings)

In this chapter we examine the experience of what it is like to live in a Passivhaus. The case studies include two developments in Germany, which are of particular interest because the residents did not come to the building as advocates of Passivhaus, although the architects did involve them in the design of their apartments and they learned about Passivhaus during the process. The residents' feedback about their homes is perhaps more typical of how most people would respond to living in a Passivhaus (as opposed to the response of a Passivhaus aficionado), provided they have some basic understanding of the building's rationale when they move in.

Common themes

There are some themes that occur in much of the feedback from residents of different Passivhaus homes. These are summarised below, before we look at responses that are specific to individual case studies.

Noise

This was something that struck many Passivhaus occupants: their home is much quieter than they were expecting. There is no omnipresent sound of a boiler churning away in the background throughout the winter. External noises are also very muted when the triple-glazed windows are closed, because the airtight construction inhibits noise transmission as well as controlling airflow. This is a big benefit for those living on or near busy main roads.

Despite the very low ambient noise levels, the ventilation system does not intrude. It is completely quiet in the bedrooms and living room, and can be heard only faintly in the bathrooms and kitchen. With it being so quiet, some people say they notice residual noise more than they would do otherwise. In a home with a lot of hard surfaces, such as wooden floors, this can make open stairwells, for example, quite noisy. Wall hangings or carpet would help in this case.

Energy bills

It goes without saying that occupants of Passivhaus buildings enjoy very low bills, but it is worth saying what a great feeling it is! You feel highly motivated to contact your energy provider to give them a meter reading. Speaking to their customer service representatives on the phone, it is hard not to feel a little smug. Knowing that the energy bills will always be manageable is in itself very liberating.

Kitchen

Conventional kitchens rely on an extractor fan over the hob to remove cooking smells. In a Passivhaus, air extraction is via the MVHR system, so extractor fans must be set up in 'recirculation mode', with charcoal filters, rather than with a direct extraction to the outside. The main kitchen extract vent is also fitted with a filter to avoid grease collecting in the extract duct. Passivhaus occupants have observed how effective the MVHR is at quietly and steadily removing steam and cooking smells – so, depending on how much frying they do, there is often little need for the recirculating charcoal filter extractor fan at all. Aside from periodic replacing or cleaning of MVHR filters, the Passivhaus/MVHR approach demands less intervention by occupants. By contrast, conventional extractor fans, which are designed to extract air very quickly, are very noisy and are not always consistently used – not least because the noise puts people off using them.

More stable temperatures in a Passivhaus kitchen is another minor benefit: for example, butter left outside the fridge stays soft enough to spread easily in winter and seldom gets too soft in summer.

Bathroom

The first benefit here is the improved thermal comfort because there are no cold walls to chill you, even if they are fully tiled. As with the kitchen, there is no noisy extractor fan of the type usually connected to the bathroom light. This is ideal when using the bathroom during the night: there is no risk of waking others in the house or of fumbling around in the dark because you don't want to use the light to avoid causing the fan to trip on.

Some occupants note the quite common use of electric towel rails. A 50W towel rail with timer and thermostat dries towels nicely and raises

Drying clothes

Drying our clothes is an activity that is quite often overlooked completely or not given the attention it deserves when homes are being designed. It is of course a design consideration in any build, Passivhaus or not. However, the MVHR system in a Passivhaus offers additional possibilities to the usual ones of opening windows and/or drying on radiators.

Ideally, it makes sense to plan for the space where clothes are dried to be near or next to the washing machine. This can be problematic in homes where the washing machine is in the kitchen, as most people do not want to dry their clothes in a relatively public space, some kitchens are too small and most have periodic cooking smells – so not an ideal environment! In a Passivhaus, one option is to build in a clothes-drying cupboard – sometimes known as a 'Swedish cupboard' – near or next to the washing machine. The cupboard is fitted with extract ventilation and a transfer path in the door or door frame so that air can easily be pulled into it. Where a home has a separate utility room with a washing machine, this room could be used for drying clothes and could therefore have extract ventilation. In a climate with colder and drier winters, where retaining humidity in the house becomes a priority to avoid overly dry indoor air, there may be an argument for placing the clothes-drying area in a space with supply ventilation; this would distribute the humidity through the rest of the house rather than it being extracted immediately to the outside.

The other possibility in a Passivhaus is to position the hot water store (see Appendix A) and the MVHR unit (and, if heating is to be provided via the ventilation system, the supply duct radiator) together with the washing machine and clothes-drying racks in a utility room with extract ventilation. This allows the warmth from the hot water store – even well-insulated hot water stores warm a room by a couple of degrees – to be used to accelerate clothes drying. (The rationale for locating the MVHR unit here, even if there isn't a supply duct radiator, is for practicality of installation and maintenance, as well as maximisation of treated floor area. If heating is provided via a supply duct radiator, this may be done using hot water: another reason for proximity to the hot water store.)

the temperature in the room. The other option is to use a minimal amount of electric underfloor matting (the electrician may need some convincing that you want only the absolute minimum (say, 200W) of matting, not the 1kW or 2kW normally fitted. Underfloor heating helps if a wetroom-style bathroom is being installed.

MVHR

Many Passivhaus occupants report that, after an initial period of tweaking and adjusting the MVHR unit (involving input from the person who commissioned the system), they very soon become unaware of it. It just does its job in the background and, being so quiet, does not impinge on your everyday consciousness. The improved air quality and lack of lingering smells and steam (from cooking or bathing) are the only things you register and that remind you that there is a ventilation system running in the background.

Arriving and leaving

Intuitively, many feel that both entry into and exit from a Passivhaus have to be rather hurried affairs because if the door is left open for more than a few seconds, it will cause the building's internal temperature to fall. Actually, this is not true. Even when people are lingering at the front door to say goodbye to guests, cold air is not drawn into the building as it would be in a leaky house. The contrast with a standard build is particularly noticeable during cold spells: one Passivhaus occupant recalled an occasion at a friend's house, when the front door was open for what was probably only five minutes, but, even though sitting at the opposite end of the house, he felt a strong, chilling draught through the whole place until the door was closed again. That doesn't happen in a Passivhaus: for there to be a draught, the incoming air needs a route out. In an airtight house, there isn't one.

Totnes Passivhaus

This retrofit project, Adam Dadeby and Erica Aslett's home in Devon, was completed in August 2011 and certified in September 2011. It was the third retrofit project in the UK to achieve full Passivhaus Certification and one of the first 20 or so of all Certified Passivhaus buildings in the UK, retrofit and new build.

Passivhaus retrofit – before and after

Being a retrofit and involving the extension of an existing house, this project provides an interesting comparison of the house as it is now with the way it was before. In the old house, there were none of the benefits we now take for granted. The building had single-glazed windows, widespread damp problems, numerous cold spots and very poor air quality. The result was an uncomfortable and unhealthy home which, despite improvised superficial improvements such as airtightness strips around windows, was a battle to keep warm. We found ourselves limiting time spent in parts of the house that were difficult to heat. We managed to keep our fuel consumption down to just above the national average, but doing so required constant attention and monitoring.

After the building work was completed, it was a different story. Some of our observations, and those of our guests, are as follows.

- It's a nice temperature and very comfortable around the house.
- Great that the bills are low.
- The house seems to take care of itself.

- You don't notice the MVHR.
- Not having radiators is good because all the wall space can be used.
- You need to go outside or have an external thermometer to know how cold it is.
- The house is very quiet; you notice the slightest sounds because there is no ambient noise.
- We expected it to be stuffy inside but were surprised how fresh and airy it turned out to be.
- It is very liberating not having to think about keeping warm or continually having to adjust thermostats or other settings.
- The house just seems to take care of you.

Temperature

The winter was quite mild, with daytime temperatures quite often reaching into double figures, and a mixture of cloudy, rainy and sunny weather: all pretty typical for wintertime in the south-west of England! During our first month in the house we had warm weather, with a couple of days in the mid- to high twenties. Unlike in some Passivhaus designs, we were forced to rely on blinds to prevent overheating in most windows that are subject to solar gain; this was because of the planners' views about the roof orientation as well as the shading constraints of the site. We quickly noticed how sensitive the temperature in the house was to our use – or non-use – of the blinds. If we used them, the house stayed comfortable and cool; if we forgot, it gained heat quite quickly. This illustrates just how important it is to design effective shading into any ultra-low-energy house approaching Passivhaus standards.

One thing that has struck us and our visitors is how stable the internal temperature is. There are no sudden fluctuations. Changes occur slowly. The only occasion on which there were persistently uncomfortable internal temperatures was during the Transition Town Totnes Open Eco-Homes Weekend,[1] when more than 150 people visited the house during a warm spell in late summer. The internal heat gains from all those bodies could not escape the building quickly enough, even with windows open. This isn't a surprise – altering a few numbers in the PHPP model of the house very quickly tells you how big an impact all that extra internally gained heat can have.

Heating system

We chose to use a supply duct radiator (pictured on page 203) to provide the very small quantity of heat needed to maintain 20°C during winter. It seems quite an elegant solution to use the ventilation system to distribute the heat as well – one system; two functions. However, we did not get the radiator connected and running until mid-January 2012. Spending a winter in a standard build without a heating system would have been very uncomfortable and inconvenient. But our temporary heating substitute, one small oil-filled electric radiator set on its lowest setting (about 400W), kept the whole three-storey house warm to about 19°C.

As autumn progressed, despite the fact that south-west England experienced an Indian summer in 2011, it was interesting to note that many homes in the neighbourhood had already started using their central heating or had started wood burning while we still felt comfortable opening our windows in November. It was good to get this confirmation of the claim that, in a Passivhaus, the 'window-opening season' starts earlier and ends later in the year than in a standard build.

During unseasonably cold weather in February 2012, there was a small but noticeable difference in temperature between the ground floor and the rest of the house. This is probably partly because of the way the ventilation is set up (as is quite common in a Passivhaus, air change rates are higher on the middle and top floors). It may also be partly due to the relatively poorer solar gain and insulation levels on the ground floor and the presence of a thermal bridge – unavoidable in a retrofit – around the base of the external wall perimeter. This means that a very small amount of additional heat (around 200W) may be needed in the main living area on the coldest winter days.

Living with the MVHR

The MVHR has been a great success. There are no lingering smells, and bathroom mirrors, if they get steamed up at all, clear quickly. Three months after moving in we changed the paper filters; a simple five-minute job, and the user control on the MVHR unit reminds you when to do it. It was surprising to see how much dirt the filters had collected during those three months. The reason the extract filter was so dirty was probably because of lingering dust in the building after all the construction work. The intake filter, which filters incoming fresh air, was also dirtier than expected, particularly considering our relatively rural location.

The ventilation system has a number of options that we haven't used to date. We just leave it on the standard setting, as one of us is at home most of the time. The unit is drawing just over 40W, equivalent to about 1kWh (a cost of about 15p) a day. During the warmer weather, it took a little while to get the settings right for when the summer bypass kicks in. The user interface on the MVHR was not so friendly in this regard, and we would hope that the challenge of designing intuitive-user controls will be addressed as MVHRs are developed further. A poorly thought-out user control can make the difference between success or failure on the part of the manufacturer in achieving the intended design goals.

Denby Dale

Denby Dale, in Yorkshire, is the UK's first Passivhaus to be built using the cavity wall construction typical of nearly all post-Second-World-War housing in the UK. Denby Dale's occupants kept a blog[2] to record their experiences during their first months in the house, from which the following text is derived. They reported similar experiences to those of others living in a Passivhaus:

- Quiet interior.
- Good indoor air quality.
- Stable indoor temperatures.

The Denby Dale house is being used as a case study by Leeds Metropolitan University and it contains a number data loggers, which record internal and external temperatures, relative humidity (RH – see Chapter 10) and CO_2 levels (see Chapter 12). Interestingly, internal temperatures have been found to be very stable over time, hardly changing during the day/night cycle, and with only small variations (typically less than 2°C) between different parts of the house. The house stayed warm during 2010/11, one of the coldest English winters in many years. During that winter the occupants experienced occasional problems with overly dry air at night-time, but they resolved this temporarily by placing a wet tea towel over a chair in the bedroom. For the green-fingered, use of pot plants would provide a more aesthetically pleasing solution. As noted in Chapter 12, it is possible to specify an MVHR unit that recovers humidity as well as heat, which is worth considering in cold, dry climates. The occupants also reported a few teething problems with their heating system (although this was not via the MVHR) and a small learning curve in getting to know how the house responds to the changing seasons.

In this house, the MVHR unit is located outside the thermal envelope. During the very cold winter of 2010/11, the unit's condensate drain was affected by frost. The condensed water from the extracted, cooled air needs to go somewhere, so this pipe has a key role in taking the water away and it is not good if it gets blocked. At Denby Dale, they remedied this by simply insulating the pipe.

Grove Cottage

Grove Cottage is a Victorian terraced house in Herefordshire, retrofitted using Passivhaus methodology, and is another pioneering ultra-low-energy project in the UK: the first to be certified as having met the EnerPHit retrofit standard. Even though this standard takes the building down to only 25kWh/m².a heat demand and 1.0ach, not the 15kWh/m².a and 0.6ach of a full Passivhaus, the occupants of Grove Cottage report many characteristics in their retrofitted home that are common to fully certified Passivhaus projects:

- Very quiet and tranquil interior because of the airtightness and triple-glazed windows.
- Very comfortable with no draughts – in huge contrast to how the house was previously.
- After some initial 'tweaking' of what is an early-model MVHR unit, it works well.
- Air quality seems very good.

Friends visiting Grove Cottage have been amazed at the lack of radiators or underfloor heating. Rather than using a supply duct radiator (which in an EnerPHit house would not be able to deliver enough heat to maintain 20°C on the very coldest days), the owners opted for the technically more orthodox approach of installing two or three conventional radiators, which run at a low temperature.

The owners enjoy being able to sit in comfort to work in the kitchen, an open area with good daylighting that before the retrofit was too cold for any use other than cooking. In their sitting room they installed a small, free-standing ethanol burner (ethanol gel burning in a pot) in an old living room hearth to create an occasional open-flame effect in winter. The burner gives off 500W of heat. The MVHR system takes care of water vapour and CO_2 released from the burner.

More generally, the householders have noticed that their home demands much less 'work' to maintain the right temperature. They commented that "the house is a comfortable 'neutral' background to family life".

Two Passivhaus apartment buildings

Two case studies that form part of doctoral research by Henrietta Lynch[3] in 2012 shed an interesting light on what it is like living in a Passivhaus apartment building. Both are in Germany. One, the KlimaSolarHaus (pictured here), a block of 19 flats in the Berlin district of Friedrichshain, has a shared MVHR system and a communal wood-pellet boiler heating system. The building was commissioned and the residents moved in during the summer of 2009. The block is run as a housing association / tenant co-op.

The other project, which was completed at the end of 2002 and was the first Passivhaus development of this type, is also a development of 19 'family-friendly' flats in a relatively central district, this time in Frankfurt, a city whose planning regulations now require that all new public housing developments are built to the Passivhaus standard. (Clearly we have a long way to go in this regard! See Chapter 14.) Each flat has its own MVHR system, but the building shares a single gas boiler.

Residents in both developments had experiences that were similar in many ways to our experience of living in the Totnes Passivhaus. In particular, they refer to:

- The comfortable indoor environment.
- Protection from external noise.
- Low energy bills.
- The lack of summertime overheating (after an initial period of learning to use the shading devices effectively).
- The affordability of the purchase price of the flats.
- The convenience of having a dedicated clothes drying space (a 'Swedish cupboard', which took advantage of the MVHR system).

In both developments there was a period in which residents had to make small adaptations, especially with regard to becoming more aware of overheating, and adjusting blinds or other shading devices accordingly. However, this was not seen as a burden. In general, the feeling was that the buildings didn't trouble them; if anything the opposite.

An important factor in the success of the projects appears to have been the involvement of the future residents in the design of the building. As well as being given what they wanted, they gained an understanding of the design process. They would also have had an opportunity to get to know each other before moving in, and this social factor would almost certainly contribute much to the success of the project.

RECAP

Although Passivhaus buildings do require a few small changes in how we 'operate' our homes, they do not demand a wholesale change in lifestyle; these are minor adjustments that, with a little prior explanation, are intuitive and very easy to accommodate into one's daily life. Worries about noise and intrusiveness from MVHR ventilation have been found to be unwarranted, and many occupants have noted the quality of the air. The comfort benefits of their homes are appreciated by many of those living in a Passivhaus.

Indeed, there is a sense of liberation provided by Passivhaus homes – not only are occupants freed from high fuel bills, but also the building just seems to 'take care' of them by maintaining a comfortable and healthy environment.

CHAPTER FOURTEEN
Policy change in the UK

Planning, a building-fabric-based energy standard, floor measurement conventions, VAT, Energy Performance Certificates, property tax, change in the construction sector, self-build, home-grown Passivhaus products, culture and policy-making

Planning

The UK planning system was established following the 1909 Housing and Town Planning Act and subsequent legislation, particularly the 1947 Town and Country Planning Act. Conceived to address the growing problems arising from the increasing density of settlement in parts of the UK, it was a key milestone and it has had a profound impact on the nature of settlement here, as well as on the look of our towns and cities. One of the main purposes of planning at local and national level is to control building development – clearly an essential role, given the scale of development over the last century. However, there are problems with the way the system currently works that makes delivering ultra-low-energy buildings harder.

Planning decision-making

The UK is unusual in the extent to which case-by-case planning decisions are made at the discretion of the local planning authority. The effect of this is to discourage smaller developers and self-builders, because most are unable to spend more than the minimum on design until after planning permission has been secured, owing to the uncertainty inherent in such a system. This impacts particularly on Passivhaus, because in order to deliver the standard as cost-effectively as possible, it is necessary to undertake more detailed design work than this minimum before submitting a planning application. Not doing so adds risk and potential cost for the house builder. Even larger developers may be unwilling to commit more than the minimum financial outlay to a potential project before it gets past the planning hurdle.

Significantly, there is a democratic role played in planning authority decisions, in that they are intended, in part, to take account of the views of local residents, as expressed in the responses to individual planning applications. And there is a conservative cultural tendency for people – whether residents, councillors or planners – to 'vote' for what they are used to. Furthermore, the planning process can inadvertently encourage verbose planning application responses that may focus on matters unrelated to the planning issues raised by the application. Indeed, some planning authorities carry warnings to the public to desist from abusing planning applicants in their responses to applications, and some choose not to publish application responses on their websites! All this consumes time and energy non-productively and feeds unpredictability and arbitrariness into planning application decision-making, despite the best efforts of those who work with integrity. It also adds to the risk that an innovative proposal will be refused planning permission, even if the applicant has sought pre-planning advice. As we saw in Chapter 4 (page 50), the planning system tends to inhibit innovative design in general.

Another problem with the current system is that planning decisions can completely undermine the objectives of what are, or should be, key national (and international) imperatives in areas as urgent and serious as climate change and energy policy. UK legislation from 1990 onwards has required local planning authorities to produce local and regional development plans. However, in practice they still seem to use their discretion in deciding to what extent the contents of these plans are taken as 'material considerations' in individual planning applications. It is possible for the plans to carry much greater weight in determining the outcome of individual planning applications than is generally the case. However, if the authorities' discretion were limited, the risk is that the planning system would become inflexible. To counter this, plans would need to be much more detailed and more frequently updated. While such development plans might cost more to maintain, public and private money would be saved by greatly streamlining the decision-making process for individual applications.

More prescriptive and detailed development plans would allow clients and architects to work within known boundaries and rules, making the undertaking of design work prior to gaining planning consent less financially risky than it is today. This change would not mean that listed buildings have to be treated the same as buildings in general – any development plan could set out exceptions for those few special buildings that are listed. However, there is an argument that too many buildings now find themselves listed or in conservation areas (or are treated by planners as though they were in conservation areas) – 5 per cent of UK housing is now in a conservation area. A review of the criteria used in determining listed or conservation status should be considered, to halt 'conservation creep'.

Planning would still be a local or regional process and as such could conflict with priorities expressed in national planning policy. However, a less discretionary planning process, more strongly driven by development plans, could link better with other government policies – and development plans could be scrutinised in the round, via a democratic process, rather than interpreted on the current application-by-application basis, where emotions can run high and personality issues can risk clouding rational decision-making. A less discretionary planning system should also be less susceptible to the risk of potential corrupt practices.

While it may be true that developing and maintaining more detailed and frequently updated development plans would potentially add to costs incurred by the State, the current planning system burdens the economy tremendously. Its capricious nature consumes a great deal of energy, time and money for planning officers, planning authority members, developers, architects, builders and clients. This takes many forms: self-censoring of designs, submission of speculative and over-sized plans as negotiating ploys, and money spent resubmitting plans following a change of heart by planners. If the cost of even some of this unproductive work could be avoided, it is unlikely that the change would represent a net cost to the economy. Clearly the cost of running the system is an important consideration, particularly given the current state of public finances.

Localism

The current UK government is promoting a policy it calls 'localism'. As with zero carbon, announced by the previous government a few years ago (see Chapter 5), it has announced a new policy goal without thinking through its implications or even providing a workable, practical definition of the objective. The stated aim of localism is to repatriate decision-making to local bodies. This means that national planning policy is being scaled back and development plans are being slimmed down. Opposition parties say that localism is a cover for public spending cuts. While this may be true, with regard to planning it is of equal concern that the localism agenda is driving a move towards an even more discretionary planning system – exactly in the opposite direction to that required. Rather than arguments emerging once in the production of a development plan, they will be repeated again and again in multiple applications. The additional costs and risks involved could act as a greater deterrent to self-builder applicants than occurs in the current system.

While some discretion will always be needed, what is required is a rebalancing, to limit the scope for discretion in relation to individual applications so that the risk to house builders can be reduced. This change would put the UK planning system more in line with many other planning systems internationally.[1] In some countries, constitutional rights further constrain arbitrariness in planning decision-making. Further streamlining could be achieved by encouraging brevity and a focus on planning criteria in application responses; this could be encouraged by the use of simple measures such as mandatory use of pro forma documents for respondents.

Planning criteria and low-energy design

In many areas, planning applications for buildings are frequently judged according to traditional material considerations:

- impact on neighbourhood or neighbours' 'amenity'
- loss of privacy or right to light
- density of development
- access, highways and similar infrastructural issues
- visual impact of materials, appearance, scale, character, setting – are they 'in keeping' with the surroundings?

This last point militates strongly against creative new ideas about the appearance, style or form of new design proposals. In the pecking order of planning criteria, a design's energy performance does not generally trump these traditional material considerations. If a design is deemed not to be 'in keeping' (a subjective determination by any reckoning), its energy performance credentials seem to gain little traction in support of an application. For example, the current weighting given to material planning considerations makes the case for external insulation on a retrofit much harder to argue successfully, because it would usually result in a significant change in the appearance of the building. While it is possible for external insulation to be faced with virtually any finish to match what was previously there, in practice it is arguable whether this improves a building's aesthetics. Also, for example, requiring a brick finish to be replicated with a false brick façade (or a genuine one) may well make it practically or financially unfeasible. The photo below shows a residential side street in Brussels. A planning system less focused on ensuring that new developments are in keeping with what already exists may result in a more higgledy-piggledy street scene such as this one. Would the world really cave in if we allowed such 'untidiness'? Indeed, should the State be the sole arbiter of what constitutes an aesthetically pleasing street scene?

Most house builders will naturally tend to act to minimise the risks associated with an uncertain planning process. They want to see the project realised, and therefore often tend to self-censor

A street view in Brussels.

their designs to avoid challenging the planning system unduly and increasing the risk that planning permission is refused. Most will also want to commission only the minimum design work needed to get through planning. This typically involves preparation of a Design and Access Statement (DAS) as well as 1:1250 block plans, 1:200 site plans and 1:100 elevations that focus on ridge heights, window positions, glazing types, building position relative to neighbouring structures, materials to be used, and wall and roof finishes.

In a Passivhaus project, factors such as the position and size of windows, and the location of fixed shading devices such as roof overhangs, have an important bearing on the building's annual heat demand, heating load and overheating risk. The design therefore needs to be entered into the PHPP, which in itself necessitates deciding on the internal floor layout (in order to determine the treated floor area); this in turn means some work on the duct layout of the MVHR system. All this is necessary in order to get meaningful energy-performance information, and means that more design work must be undertaken and costs incurred at risk before planning permission is gained – even more so in a retrofit project or with an unusual design, or where the architect or Passivhaus Designer are new to Passivhaus methodology. Unless the client is strongly committed to the goals of the project and financially able to take the risk, the planning process will act as a deterrent factor, slowing growth in the self-build sector and in Passivhaus and ultra-low-energy construction in particular. Conversely, if a planning application for a Passivhaus project is submitted without sufficient design and energy modelling work, you will be locking yourself into a design that is potentially a lot less efficient. This will probably result in more expense than is necessary to achieve the Passivhaus standard. Even where the planning system doesn't deter to the extent that a 'difficult' project fails to get off the ground, design choices will still be influenced by the wish to minimise planning risk.

In the Totnes Passivhaus project, more east–west glazing was used than is ideal in a Passivhaus, mostly to make up for restricted solar gain on the south façade, as well as client preference for good daylighting in key living areas. Unlike south façades, where fixed overhang shading devices provide effective summertime shading because the sun is high in the sky in the middle of the day, east–west windows need shading in front of the window (often external blinds) to keep out morning and late-afternoon solar gain when the sun is low in the sky. In many parts of Europe, external shading using Venetian blinds or roller shutters is quite common. However, this is not part of the UK's architectural tradition, and so to avoid prejudicing the planning application, windows with integral blinds, such as the one pictured opposite, were chosen: the external blinds are contained within an additional pane of glass, giving the windows a conventional appearance that was less likely to 'offend' the planners.

In a planning system with strengthened development plans, although local planning authorities would produce these plans, central government could establish a binding framework of national policy objectives that the regional or local plans had to address (or at least not conflict with). While some of these may relate to broad standards, it could also stipulate specific matters and how these should be weighted. In relation to low-energy design, these would need to include the following:

- Impact of form factor on energy performance and cost. House builders should not be pressurised by the planning system to design buildings with an inefficient form factor.

Chapter Fourteen • Policy change in the UK 221

A window with integral blinds (external and internal views).

- Acceptance of the use of external shading devices (e.g. blinds, shutters and overhang eaves).
- Acceptance of external insulation by default (e.g. in retrofits, greater flexibility about brick or stone finishes being replaced with insulated, rendered finishes), unless a listed building. A more flexible approach should be used in conservation areas.
- External insulation (projecting beyond the wall and roof lines of neighbouring buildings)
- Window design (performance versus traditional window aesthetics, e.g. use of mullions). While ultra-low-energy windows with mullions and muntins exist, they cost significantly more; the circumstances in which these are required by the planning system should be more limited.
- Significance of having openable windows. Adequate natural ventilation is vital in a Passivhaus.
- Roof orientation and use of projecting angled frames for solar panels to allow better optimisation of renewable energy generation. While not a Passivhaus-specific issue, this type of change would make planning decisions align with national energy policy.

Planning professionals need a properly coordinated and adequately funded programme of 'continuing professional development' training to help them gain a better understanding of how their decisions impact on the energy performance and construction costs of buildings.

Currently, these factors, in so far as they are understood at all, are seen by some in the planning system as falling within the rubric of 'sustainability' and, as such, represent merely the demands of one of a number of competing sectional and (in their own terms) equally valid interests. In other European countries, there is a broader recognition across society that there is a common interest (economic, social and environmental) in cost-effective low-energy building. This partly explains why in Frankfurt in Germany, for example, all new public housing developments must be built to Passivhaus standards.

The bigger picture

While we all have our opinions of what is architecturally 'in keeping', our energy and climate predicaments must surely be treated as overriding collective concerns. If we fail to address these most profound challenges intelligently, other issues that we care about will be subsumed by multiple crises on a far greater scale. Reducing fossil-fuel energy consumption in our housing stock is one of the easier ways to reduce energy use and carbon emissions without impacting on 'lifestyle'. By comparison, reducing fossil-fuel use in transport is much harder without big changes in societies dominated by private transport. If we are unable to reduce the demand for energy in our housing stock, we will be forced to save it in other more difficult areas.

In facing those challenges, our planning system will continue to have a vital role; not just in ensuring that it does not put unnecessary obstacles in the way of delivering ultra-low-energy buildings. Even more fundamental is its role in protecting our agricultural land so that we will be able to feed ourselves on a much leaner-energy diet. Planning is one of many policy areas that need to adapt to reflect our changing energy and environmental reality.

A building-fabric-based energy standard

As has been observed elsewhere in this book (see Chapter 5), within the UK there is currently no statutory, ultra-low-energy whole-building energy standard based on how much energy a building uses before taking account of any add-on technology such as solar panels. The result is that we are constructing buildings that are comparatively inefficient. Only with the help of expensive technology with a short lifespan relative to the building itself (e.g. wood burners or boilers, or solar photovoltaic systems) is it possible to reduce or offset the carbon emissions from their still-considerable energy use, in order to be defined as low carbon or zero carbon. This relatively complex technology needs to be maintained and correctly operated, and any energy assessment of the building that relies on it is valid only for the working life of the technology. There is nothing to stop a wood-chip boiler, for example, being replaced with a natural gas or multi-fuel one at a later

kWh versus carbon emissions

If we are to respond rationally to the 'bigger picture' challenges (see box, left), we need meaningful, effective indicators (measures) of progress towards our goals. As consumers of energy, our primary goal must be to reduce our burden on the energy generating system by demanding much less energy and using it to much greater effect. This will help those responsible for generating energy to 'decarbonise' energy production by giving them a less onerous task. We receive energy bills that use kWh (kilowatt hours) as energy units. Any meaningful measure of how efficient a building is should, therefore, be based on these units rather than on the more remote (in everyday terms) concept of carbon emissions.

date, thereby undermining the original low-carbon design intent.

An energy-efficiency standard based on the performance of the building fabric, together with other measures to help the construction sector to adapt, would allow the UK to reduce energy demand arising from its building stock. The latest version of the Code for Sustainable Homes (CSH) has started to address this by the introduction of a Fabric Energy Efficiency Standard (FEES) – an encouraging sign that Passivhaus and zero carbon are becoming more closely aligned (see Chapter 5).

While the Passivhaus standard is very demanding, Passivhaus methodology could form the basis of a set of building-fabric energy standards. (This approach has been developed by the AECB with its CarbonLite programme.[2]) A suggested set of energy-performance standards for typical residential and non-residential buildings is given in Table 14.1 below.

A convention for measuring floor area

The unit of measure for the above suggested standards – kWh/m².a (kilowatt hours per square metre per annum) – is based on floor area. A prerequisite for such standards, therefore, is an agreed statutory convention for measuring floor area, used by all relevant professional groups. Without this, there is no way to compare one building with another accurately. However, such a single convention does not yet exist in the UK. As we saw in Chapter 7 (page 94), Passivhaus uses a very specific definition of what constitutes treated floor area (TFA). For obvious reasons, the Passivhaus Institut chose to use the pre-existing German regulation as the basis for this definition. Ideally, we in the UK would adopt a common standard with other European countries, or (more challenging still) an international standard, allowing the UK to accurately compare its performance with those of other countries. This would help drive improvements globally.

VAT as a lever for best building practice

Currently in the UK, a lower 5 per cent VAT rate is applied to certain items – e.g. some insulation and solar installations – as long as they are fitted by a builder or installer with the correct registration (e.g. MCS for solar). Otherwise, UK law currently allows developers and self-builders on

Table 14.1 **Suggested energy standards for the UK**

Standard	Annual heating demand (kWh/m².a)	Airtightness standard (ach)
'A+' (Passivhaus – new builds)	15	0.6
'A' (EnerPHit – retrofits)	25	0.6/1.0
'B' (equivalent to AECB Silver Standard)	40	1.5
'C' (approx. equivalent to proposed zero carbon FEES target)	40-50*	3.0

* Adjusted to reflect Passivhaus floor area measuring convention.

new builds to claim back all the VAT paid on the construction costs, but not professional fees. Retrofits and extensions are subject to full VAT, currently 20 per cent. This represents a massive financial disincentive to retrofitting existing properties.

This skewed policy could be replaced by a new VAT regime that is neutral in its treatment of new builds and retrofits while being revenue-neutral in tax terms. Even better, why not use the VAT system to encourage an improved FEES? All that would be needed is a reasonably accurate system of assessing the energy performance of buildings.

Energy Performance Certificates (EPCs)

Within the EU, buildings are being assessed for their energy performance – every property is required to be assessed once every ten years. Under the UK's interpretation of the EU Directive,[3] properties have to be assessed much more frequently: when new, between tenancies and every time the property changes hands. The assessments are known as Energy Performance Certificates (EPCs), an example of which is shown on the right. Perhaps in order to cut the cost of these overly frequent assessments, or perhaps because the UK left it too late to put timely and well-considered arrangements in place, EPCs are based on a simplified form or 'reduced data set' version of the Standard Assessment Procedure (SAP). Unfortunately, this has meant that EPCs have been of very limited success in providing a meaningful measure of a building's energy performance.

A reformed EPC, focused on the building-fabric performance, could provide the information needed to determine the applicable rate of VAT

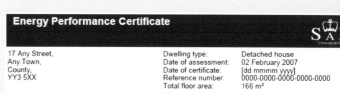

An Energy Performance Certificate.

on all retrofits and new builds. The VAT regime should be designed to make it a disincentive for house builders to construct energy-hungry structures. Where recent building work has been completed, the EPC could encompass broader measures, such as the environmental impact of the building materials used – provided, of course, that a well-thought-out system of measuring these additional criteria had been formulated.

Property tax

With its housing stock in such poor general condition, particularly in relation to energy performance, the UK has a challenge to protect poorer people from fuel poverty. The ongoing squeeze on credit since 2008, and low levels of personal saving in the UK, have restricted access to the 'property ladder' for many would-be homeowners. The result has been a big increase in the number of people renting. The problem of fuel poverty is made worse for people renting in the private sector, as they are not able to make any but the lowest-cost, least-invasive energy performance improvements to their homes. In a climate where demand for rented property outstrips supply in many areas, the tenant has even less power; if tenants complain too much, there are always plenty more to replace them.

Until 1990 (1989 in Scotland), residential properties were subject to a property tax – rates – based on the rateable value of the building in question. Local rates were chargeable to the property owner, even if the property was rented out to tenants. After rates were replaced first by the poll tax ('community charge') and swiftly thereafter by council tax, property tax became payable by residential tenants in rented properties, not by their landlords. Business premises are still subject to rates, which are payable by the tenant.

There has not been much success in reforming local property tax in the UK. Perhaps the trauma of the poll tax experience has contributed to policy paralysis. However, a reformed property tax, weighted according to a building's EPC rating as well as or instead of its value, and payable by the owner, not the tenant, could be used as a lever to send an unambiguous signal that makes owning energy-guzzling buildings economically unattractive. This area of policy would need careful consideration, as many rental properties are in communal buildings, e.g. a flat in a block of flats, and landlords are therefore not always in a position to make energy-performance improvements.

Supporting change in the construction sector

The UK government has not yet chosen to invest in retraining the construction sector so that it is properly equipped to deliver successful, cost-effective low-energy builds. The experience of the last few years has shown that driving change solely via regulation has taken the UK down the path of techno-fix solutions and a box-ticking approach rather than an effective focus on getting the building fabric right. Regulatory change needs to be accompanied by practical training support, as was the case in Canada's implementation of their 'Super E' standard (see Chapter 4, page 49).

As has been discussed elsewhere in this book, construction workers on an ultra-low-energy build need to understand the role that airtightness, thermal bridging and ventilation play in achieving very low energy use. Vocational training for the building trades needs to include an overview of building physics, including moisture management, and cover the wider social and economic effects of poor building energy performance. It is not enough that students learn how to achieve good airtightness or to avoid thermal bridging; they need to know why they are being asked to pay such close attention to what 'common sense' might suggest is unimportant. As noted in Chapter 4, the UK's education system has not been good at combining academic study and vocational training, and opinions differ about how much weight should be given to academic education versus vocational training. But it seems daft to assume that, just because an individual is not achieving enough academically, he or she should be encouraged to direct their efforts solely to vocational training. A relatively immobile class structure and old-fashioned attitudes to class still fuels a tendency to pigeonhole people as 'blue collar' or 'white collar'. There are benefits to a more holistic education and training system that encourages us to develop both vocationally and academically.

From outsourcing to 'insourcing'!

Outsourcing came into fashion in political and management thinking during the 1980s. Companies were encouraged to focus on 'core competence', shedding workers whose roles fell outside their core business. Flexible working practices were encouraged to make companies more nimble; more able to adapt to fluctuating requirements. The construction sector is more subject to the cyclical swings in the economy than most, so outsourcing helps construction firms big and small to cope.

The boom-and-bust swings in the construction sector have been made worse in some countries because they have become so dependent on property speculation to fuel their economies. Post-2008, the ongoing squeeze on credit and energy supplies is making a return to pre-2008 levels of speculative building much less likely. This means that the UK's construction sector is likely to continue to be dominated by the practice of outsourcing.

Building any structure with multiple services and a weather-proof envelope requires good communication and cooperation between all the operatives on-site; in the case of a Passivhaus build, good communication is essential. On so many projects, however, with atomised groups of outsourced workers, sometimes complete strangers, the vital communication and cooperation needed often does not take place. Contractors are tasked and motivated to deliver only their package of work in isolation, without reference to the potential knock-on effects their work could have on others. Many outsourced workers never even see the completed building and therefore have no investment in the overall goals of the project. As we have seen throughout this book, this way of working is quite antithetical to what is needed in a Passivhaus build.

A reduction in the use of outsourcing would not only put into place a framework that encourages less transient teams of site operatives but also bring wider benefits: construction workers would enjoy greater job security and satisfaction, and there would be a shift away from the corrosive blame culture that permeates so much of the construction sector. It is demotivating to work in an environment where so much personal energy is consumed in working defensively to avoid being blamed for errors and omissions. This generally arises from gaps and overlaps in the responsibilities of different trades. Most construction workers (indeed most workers in any sector) dislike this type of working environment, and this in itself is reason enough to encourage long-term multidisciplinary teams within the construction sector.

Clearly, moving away from today's extreme levels of outsourcing will be a challenging task, and will require State intervention in some form – we have to find a way to make ultra-low-energy building cost-effective and widespread. Other countries are managing to do this; are we to say that our best response is defeatist hand-wringing? Perhaps moving back to smaller, 'human-scale' construction companies would facilitate stronger team working, even if some outsourcing continues.

Promoting self-build

'Self-build' refers to any building project that is commissioned by the people who are going to live in the property. The majority of self-builds are constructed by professional contractors, but some are built partly or mostly by the prospective owners. Although self-build has enjoyed growth in recent years, it still represents only about 12 per cent of all new housing in the UK. This contrasts with upwards of 50 per cent across Europe. In countries such as Austria, self-builders are responsible for up to 80 per cent of new houses. The self-build sector is also stronger in the USA. The predominance of PLCs (public limited companies) in UK house building has encouraged unadventurous building to minimum energy-performance standards: as discussed in Chapter 4, large housing development firms have evolved building methods or formulae that deliver saleable homes for a reliable cost; they have no interest in building to reduce their purchasers' running costs. Where customers express an interest in lower bills, the housing developers have to manage this market expectation in ways that do not threaten their shareholders' interest in a reliable return on their investment. Unless the status quo is losing them money or made illegal, they will be loath to change, as to do so brings cost risks.

Self-builders have an incentive to invest up-front in order to make their home cheaper to run. They also tend to be more willing to innovate, perhaps because they are less averse to risk (or less aware of the financial implications of some of their choices). Or it may simply be that the demands of their mortgage providers are less burdensome than those of the large housing developers' shareholders. The biggest obstacle self-builders face is lack of access to building plots. Britain is unusual in the extent to which land ownership is concentrated in the hands of the very wealthy few families, corporations and State agencies: the Royal Family, the Ministry of Defence, the Forestry Commission, local authorities, the aristocracy and, more recently, ultra-rich overseas purchasers. According to Kevin Cahill, writing in 2001,[4] less than one per cent of the population owned more than two-thirds of land in Britain, although clearly not all of this is suitable or appropriate for housing development. The Scottish Parliament passed a Land Reform Act in 2003 that gave Scots defined rights to buy limited land without the owner's consent. However, when land is released by the planning system for building development, it still almost invariably goes to the large housing developers. The problem of access to land is made even worse by the practice of 'land banking' by developers and specialist land banking companies: land on the edge of towns is bought up and held in anticipation of a future change in planning status or higher future house prices. This state of affairs limits self-builder innovators to the relatively wealthy, who can afford to buy the few expensive available plots. Even the more affluent self-builders are often forced to buy an existing property for retrofitting or for demolition and rebuild; itself not a cheap option.

A modest land tax would provide a disincentive to unproductive hoarding of land. Provision is also needed to ensure that much more of the land released by planning authorities for building is given over to self-builders and small-scale developers. In this respect, an encouraging development is to be found in the latest draft National Planning Policy Framework (July 2011), which includes the proposal that an obligation should be placed on local councils to meet the self-build demand (see Chapter 4, page 48).

Encouraging home-grown Passivhaus products

Currently, the countries that lead the world with Passivhaus, Austria and Germany, also lead the world with the manufacture of products that help you build to Passivhaus standards. The UK needs to make the capital costs of building a Passivhaus as competitive as possible. We also have to address the need to reduce supply chains so that a greater proportion of building materials is produced nearer to their point of use, particularly those that are heavy or bulky to transport and that are not complex to make. In a region the size of the UK, there is also no reason why more specialist products, such as Passivhaus-certified triple-glazed windows or heat recovery ventilation systems, could not be made domestically. The aim should be to make these products mainstream, just as double glazing is now.

The UK government does not have a track record of providing incentives to this sort of productive sector, which, if it were as healthy as its German or Austrian counterparts, would make a significant contribution to the economy and to achieving the country's social and environmental goals. The Austrian province of Vorarlberg has been very successful in bringing together all the policy levers necessary to reduce energy demand from housing and promote the development of suitable domestically manufactured products. The UK government expresses reluctance to interfere in the 'free market' economy, but this stance is very selective, as its frequent interventions in its favoured sectors – car building or, most spectacularly in recent years, the finance sector – have demonstrated. In contrast to this position, there is an economic argument that in capitalist economies new sectors will not grow without initial support from the State until they reach self-sufficiency.

Culture, politics and the policy-making process

Since 1979, when neoliberal economics took hold of government thinking in the UK, the arguments have been that 'the State shouldn't intervene' or 'the State isn't competent to intervene'. These ideas have become embedded in much of the UK's mainstream political discourse in a way that is not the case for most of our Continental neighbours. Even during the Labour years of 1997-2010, the nature and extent of State intervention was constrained by the pervasiveness of this notion.

In parallel with, or perhaps pre-dating, this loss of belief in the efficacy of the State has been the growth of a culture of short-termism – both in policy-making circles and more broadly. It has had many impacts: as discussed in this book, house building is subject to its effects. From the late 1980s until recently, both Conservative and Labour governments embraced the private finance initiative (PFI) as a means of funding new capital projects in the public sector without adding to the public sector borrowing requirement – real something-for-nothing financing. We also see this tendency in retailing: subscription-based products, such as home phone or Internet services, are marketed with the first few months at half price; bubble-jet printers are sold very cheaply but the ink supplies are very expensive. These marketing strategies work because their customers are solely focused on up-front costs. Even PLCs, by focusing relentlessly on maximising shareholder value to the exclusion of almost all other objectives, have created incredible short-termism in the corporate culture. Similarly, the finance sector, which used to have a role supporting the long-term growth of the small- and medium-sized enterprise (SME) sector, has given this up in favour of ultra-short-termism – 'casino capitalism'.

Added to this, in more and more areas, the government's role in policy-making is being pared down to that of a cipher for lobbyists representing the interests of large corporate groupings. An example is role of the Home Builders' Federation (HBF) in co-chairing the government's '2016 Taskforce',[5] which is addressing barriers to achieving the zero carbon standard. While it is makes sense for the HBF to have an input, the government needs to be the one in the driving seat, setting the framework and critically appraising the submissions of the HBF and other interested parties, such as the National House Building Council (NHBC). As we have seen in this chapter, moving towards a Passivhaus future, whether or not it includes zero carbon, will require difficult changes, and HBF members may not regard these as being in their (short-term) interest.

In a world of Twitter and non-stop rolling news, the media has fed the culture of short-termism. In the compressed timescales of today's news-gathering, it has become virtually impossible to convey complex issues, particularly when understanding them requires an explanation of the science and the numbers. The demands of the media for catchy headlines and instant diagnoses of problems, swiftly followed by 'solutions', have reduced the quality of debate and encouraged governments to pass poorly thought-out legislation in haste.

In the face of all these trends, it is not surprising that the process for developing policy has been so undermined. Making Passivhaus happen on a large scale in the UK requires a sea change in the way we create and implement policy, and a renewed belief in the ability of State agencies to pull together all the elements of policy needed to achieve a desired objective. That change will not happen unless we have the courage and imagination to do things differently, at all levels.

RECAP

There is much scope in the UK for changes in public policy – and, crucially, in the public-policy-making process – that would help the UK realise its ambition to construct all new housing to a very high energy-performance standard, and to do so cost-effectively. The areas where such changes are required range from planning to the provision of training and support to develop new skills and knowledge in the construction sector. Of course, it will take time for such changes to be conceived and implemented properly, and before the benefits become apparent. The key areas where change is needed are as follows.

- Planning – a reform of the system, strengthening the role of development plans and putting limits on the scope of discretionary application-by-application decision-making.
- Development of a fabric-based energy standard.
- Development of an agreed convention for measuring floor area.
- Changes to the VAT regime as it applies to construction.
- Changes to property and land tax, and land disposal.
- State-supported training to help the construction sector adapt to low-energy building.
- Promotion of self-build.
- Encouragement to promote home-grown Passivhaus products.

The policy-making process needs to avoid the pursuit of short-term political gains and must be driven by government without the undue influence of corporate lobbyists.

Appendix A
Space heating and hot water

Space heating and hot water options in a Passivhaus (aside from MVHR)

Fuel	Advantages	Disadvantages
Electricity (direct, e.g. electric radiator or immersion heater)	<ul><li>Very low capital cost</li><li>Controllable and convenient</li><li>Can also be used to heat water separately from space heating</li></ul>	<ul><li>High carbon. Electricity is therefore considered a poor choice as the main source of space heating or domestic hot water heating</li></ul>
Heat pump (ground or air source)	<ul><li>If designed and installed correctly, similar carbon emissions to natural gas</li><li>Lower maintenance – no boiler servicing</li></ul>	<ul><li>Knowledge of how to design and install heat pumps is not widespread</li><li>High capital cost</li><li>Many systems are suitable only for heating to low temperatures and do not work efficiently when heating domestic hot water to 60°C, so could only be used economically for space heating</li></ul>
Liquefied petroleum gas (LPG) / oil	<ul><li>Similar advantages and disadvantages to natural gas (see below), but with higher carbon emissions</li></ul>	
Natural-gas-fired boiler	<ul><li>Widespread suppliers/knowledge, low(ish)-cost technology</li><li>Lower carbon than oil or grid electricity</li><li>Efficient, controllable and convenient</li><li>Can also be used to heat water separately from space heating</li></ul>	<ul><li>Very hard to source boilers that can reduce ('modulate') their output low enough for a typical single-family-dwelling Passivhaus*</li><li>Relatively complex technology</li><li>Carbon intensity of natural gas is increasing as its EROEI declines (see Chapter 5)</li><li>Requires routine servicing and (if property is occupied by tenants) annual gas safety certificate</li></ul>
Wood burner (logs)	<ul><li>Low carbon and a simple technology</li><li>Good solution in rural areas and in larger buildings (i.e. those with larger heating needs)</li></ul>	<ul><li>Hard to source wood burners with low enough output for space heating*</li><li>Less convenient – requires daily attention</li><li>Less suitable in multiple-dwelling buildings in towns and cities</li><li>If a wood burner is also used to provide hot water, even with solar thermal panels for the summer months, during the shoulder months of the heating season no space heating is needed but water heating is. This can be solved by using immersion (direct electric) heating for hot water during these periods</li></ul>
Woodchip-fired boiler	<ul><li>Low carbon; an alternative where there is no mains natural gas</li><li>Woodchip boilers more tolerant of variations in the properties of woodchips than pellet boilers are of pellets</li><li>Can also be used to heat water separately from space heating</li></ul>	<ul><li>Very hard to source boilers that can reduce ('modulate') their output low enough for a Passivhaus*</li><li>Reliant on a limited number of specialists for correct installation and maintenance of boiler</li><li>Less efficient than wood-pellet-fired boiler</li><li>Space needed to store woodchips (hopper)</li><li>High capital cost</li></ul>

(Cont.)

Fuel	Advantages	Disadvantages
Wood-pellet-fired boiler	<ul><li>Low carbon; an alternative where there is no mains natural gas</li><li>As efficient, controllable and convenient as mains natural gas</li><li>Can also be used to heat water separately from space</li></ul>	<ul><li>Very hard to source boilers that can reduce ('modulate') their output low enough for a Passivhaus*</li><li>Reliant on a limited number of specialists for correct installation and maintenance of boiler and supply of pellets. The boiler is dependent on the right specification of pellets.</li><li>Space needed to store wood pellets (hopper)</li><li>High capital cost</li></ul>

* This situation is changing as manufacturers respond to the increased demand for products with the very low outputs needed in ultra-low-energy buildings.

A note on hot water storage

In the UK, it is common to use 'combi' gas boilers, which provide instant hot water and do not require any hot water storage (although better combi boilers do now sometimes include a small integrated water store to avoid the boiler firing up every time hot water, however little, is used). Combi boilers are popular because they avoid the expense of a separate, full-sized hot water store and also save valuable floor space (useful in the relatively cramped space of much UK housing).

Use of combi boilers or other instant hot water solutions makes it impossible to use solar thermal panels, as solar panels are not able to reliably deliver heat at a consistent and sufficiently fast rate to heat water as it is being used. While Passivhaus does not specify whether to use an instant hot water system or a system based on a hot water store, it does encourage the use of solar thermal panels, where the building's orientation and site are suitable; this solution, then, would require a hot water store.

Thermal stores

A **thermal store** provides the same function as a conventional hot water tank or store. But, unlike in a hot water tank, where the hot water is drawn out and used directly in the taps, the hot water in a thermal store is never used directly. Instead, cold mains water is fed through a coil in the thermal store, takes in heat from the surrounding water, and emerges hot. Thermal stores have two advantages over conventional hot water tanks. Firstly, they can work at a lower operating temperature. In a conventional hot water tank the temperature must reach 60°C at least once a week to protect against legionnaires' disease. For this reason, in the UK they are generally set to maintain this temperature permanently. Current regulations require that hot water coming out of the tap must not exceed 48°C to avoid the risk of scalding, so hot water from the tank is mixed down with cold. A thermal store can be set to maintain a minimum temperature of, say, 50°C. This lower temperature reduces the small but unwanted heat gains in a Passivhaus and, if solar hot water panels are used, means the secondary source of heat (e.g. a boiler) doesn't have to raise the water temperature so much. Secondly, because the water in them is not under high pressure, thermal stores can be connected to 'uncontrolled' heat sources, such as the back boiler of a wood burner, making them a more flexible option should logs be used to provide water heating, either now or in the future. The lower temperature also makes the use of heat pump technology easier, if this is your choice of heating method. (Heat pumps work more efficiently when the temperature difference between water in and water out is smaller, in the region of 35°C.)

Appendix B
Thermal conductivity values

Lambda 90/90 values (λ90/90) are the values adopted in the UK for Passivhaus calculations. These are thermal conductivity values that have been calculated according to the Lambda 90/90 convention (see box on page 96). This is normally signified by the CE mark.

The values given in the table below are not Lambda 90/90 values, as such figures can relate only to specific products rather than to generic materials. Some of the figures given here are approximate, while some vary according to the temperature and other properties of the material.

Conductivities for common building materials/elements

Structural materials	Lambda value (λ), in W/mK	Insulation	Lambda value (λ), in W/mK
Oriented strand board (OSB)	0.13	Phenolic foam	0.021-0.024
Softwood (across grain)	0.13	Polyisocyanurate foam (PIR)	0.022-0.028
Porous/lightweight concrete	0.15-03	Polyurethane (PUR)	0.022-0.028
Hardwood (across grain)	0.18	Extruded polystyrene (XPS)	0.029-0.039
Gypsum plaster	0.18-0.56	Expanded polystyrene (EPS)	0.030-0.038
Gypsum plasterboard	0.25	Glass mineral wool	0.031-0.044
Brickwork (typical)	0.6-1.2	Rock mineral wool	0.034-0.042
Stonework	1.5-1.8	Sheep's wool	0.035-0.04
Reinforced concrete	2.1	Cork insulation	0.036-0.06 (typical)
Stainless steel	16	Cellulose (e.g. Warmcel®)	0.038
Carbon steel	43	Woodfibre boards (density-dependent)	0.038-0.047 (typical)
Cast iron	55	Hemp bats	0.04
Aluminium	250	Cellular glass (e.g. Foamglas®)	0.041-0.06
		Strawbale*	0.055-0.082
		Calsitherm® Climate Board (calcium silicate, high capillary action)	0.059
		Tradical® Hemcrete® (hemp insulation)	0.06

* In load-bearing strawbale buildings this acts as both a structural material and insulation.

Appendix C
US units – metric conversions

US units	Metric units
Energy	
1 BTU (British thermal unit)	0.000293kWh (0.293Wh)
1 therm	29.3kWh
Specific energy consumption (e.g. to measure annual specific space heat demand)	
1 BTU/ft².a (BTUs per square foot per annum)	1 kWh/m².a
Power	
1 BTU/hour	0.293W
Specific power (to measure heat or cooling load)	
1 BTU/hr/ft² (BTUs per hour per square foot)	3.1538W/m²
Temperature	
Fahrenheit	°F = (°C x 1.8) + 32
Length	
1 inch	25.4mm (0.0254m)
1 foot	0.3048m
Area	
1ft² (square foot)	0.0929m²
Volume	
1 US gallon	3.7854 litres
1ft³ (cubic foot)	0.0283m³
Volume flow (to measure ventilation rates)	
1 CFM (cubic foot per minute)	1.699m²/hr
Thermal conductivity	
1 BTU/(ft.hr.F) (BTUs per foot per hour per degree Fahrenheit)	1.7304W/mK
U-value ('U-factor')	
1 BTU/(hr.ft².F) (BTUs per hour per square foot per degree Fahrenheit)	5.6769W/m²K
Electrical efficiency (of MVHR units)	
1.308 CFM/W (cubic foot per minute per watt)*	0.45 Wh/m³ (watt hours per cubic metre)

* Note that this unit is the inverse of the metric unit.

Appendix D
Certified Passivhaus projects in the UK*

Project	Description	Image
Camden Passivhaus by bere:architects	Single family house, private client. CAMDEN, NORTH LONDON Treated floor area: 100m² Airtightness: 0.4ach Annual space heat demand: 13.3kWh/m².a Primary energy demand: 91kWh/m².a	Image: Jefferson Smith
Canolfan Hyddgen Training Centre by JPW Construction	Offices, training and education centre for Powys County Council. First Passivhaus building in the UK. Also BREEAM 'Excellent' standard. MACHYNLLETH, POWYS, WALES Floor area (unspecified): 410m² Airtightness: 0.25ach Annual space heat demand: not known Primary energy demand: not known	Image: JPW Construction
Carnegie Village, Leeds Metropolitan University Headingley Campus by GWP Architecture Ltd	Student accommodation (479 study bedrooms). BREEAM 'Excellent' standard; also 2010 BREEAM Multi-residential Award. LEEDS Treated floor area: not known Airtightness: not known Annual space heat demand: 14kWh/m².a Primary energy demand: 63kWh/m².a	
Centre for Disability Studies by Simmonds.Mills	New-build offices and meeting rooms for a charitable organisation. BREEAM 'Excellent' standard. ROCHFORD, ESSEX Treated floor area: 307m² Airtightness: 0.33ach Annual space heat demand: 15kWh/m².a Primary energy demand: 81kWh/m².a	

(Cont.)

* To end of December 2011. Original list and images as compiled by A. J. Footprint in conjunction with UK Passivhaus Trust, with added detail. For updated lists visit www.passivhaustrust.org.uk/projects.

Appendix D

Project	Description	Image
Crossway by Hawkes Architecture and consultants Scottish Passive House Centre (SPHC)	Single family house, as seen on *Grand Designs*, private client. STAPLEHURST, KENT Treated floor area: 285m² Airtightness: 0.56ach Annual space heat demand: 15kWh/m².a Primary energy demand: 61kWh/m².a	*Image: Richard Hawkes*
Denby Dale Passivhaus by Green Building Store	Single family house, private client. DENBY DALE, YORKSHIRE Treated floor area: 104m² Airtightness: 0.3ach Annual space heat demand: 15kWh/m².a Primary energy demand: 87kWh/m².a	*Image: Green Building Store*
Dormont Estate by White Hill Design Studio and consultants SPHC	Eight semi-detached houses for commercial rent – each includes log-burning stove with back boiler. LOCKERBIE, DUMFRIES & GALLOWAY, SCOTLAND (through Scottish Rural Affordable Homes for Rent pilot scheme) Treated floor area: not known Airtightness: 0.6ach Annual space heat demand: not known Primary energy demand 6430kWh/a	
Grove Cottage by Simmonds Mills Architects	Retrofit (EnerPHit) single family house, private client. HEREFORD Treated floor area: 135m² Airtightness: 0.79ach Annual space heat demand: 25kWh/m².a Primary energy demand: 102kWh/m².a	
High Barn by Hanse Haus	Prefabricated four-bedroom detached house, private client. WEST QUANTOXHEAD, SOMERSET Floor area (unspecified): 250m² Airtightness: not known Annual space heat demand: 13kWh/m².a Primary energy demand: 30kWh/m².a	

(Cont.)

Project	Description	Image
Kingdom House by Oliver & Robb Architects	Social housing unit for Kingdom Housing Association. BREEAM EcoHomes 'Very Good' standard. PITTENWEEN, FIFE, SCOTLAND Gross floor area: 104m² Airtightness: 0.58ach Annual space heat demand: 14kWh/m².a Primary energy demand: 85kWh/m².a	
The Larch House by bere:architects	Affordable housing, United Welsh Housing Association. Zero carbon – CSH Level 6. EBBW VALE, BLAENAU GWENT, WALES Treated floor area: 86.7m² Airtightness: 0.2ach Annual space heat demand: 13kWh/m².a Primary energy demand: 83kWh/m².a	
Lena Gardens by Green Tomato Energy	Victorian house retrofit for private client in conservation area. Solid brick walls internally insulated. HAMMERSMITH, WEST LONDON Treated floor area: 195m² Airtightness: 0.49ach Annual space heat demand: 12kWh/m².a Primary energy demand: 69kWh/m².a	
The Lime House by bere:architects	Affordable two-bedroomed house for United Welsh Housing Association. CSH Level 5. EBBW VALE, BLAENAU GWENT, WALES Floor area (unspecified): 76m² Airtightness: 0.47ach Annual space heat demand: not known Heating load: 10W/m² Primary energy demand: 88kWh/m².a	
Mayville Community Centre by bere:architects	Retrofit of nineteenth-century brick building to create new community centre. ISLINGTON, NORTH LONDON Treated floor area: 665m² Airtightness: 0.43ach Annual space heat demand: 11kWh/m².a Primary energy demand: 127kWh/m².a	

(Cont.)

Appendix D **237**

Project	Description
Nash Terrace by 4orm Architects	Four terraced town houses, solid timber structural panels, private developer. HIGHBURY, NORTH LONDON Floor area (unspecified): 292m² Airtightness: 0.4-0.6ach Annual space heat demand: 11kWh/m².a Primary energy demand: 95kWh/m².a
Princedale Road by Paul Davis + Partners, and Philip Profitt (Princedale EcoHaus Ltd)	Victorian terrace retrofit in conservation area for Octavia Housing Association. Internally insulated. First Passivhaus-certified retrofit in the UK. NOTTING HILL, WEST LONDON Treated floor area: 88m² Airtightness: 0.5ach Annual space heat demand: 11kWh/m².a Primary energy demand: 62kWh/m².a
Racecourse Estate by Mark Siddall of Devereux Architects	Four terraced Passivhaus bungalows for Gentoo Homes. CSH Level 4 without renewable energy sources. HOUGHTON-LE-SPRING, CO. DURHAM Treated floor area: 67m² Airtightness: 0.58ach Annual space heat demand: 15kWh/m².a Primary energy demand: 97kWh/m².a
Sampson Close by Baily Garner	Social housing for Orbit Heart of England – mix of flats and terraced housing. COVENTRY Treated floor area: not known (mixed units) Airtightness: 0.6ach Annual space heat demand: 15kWh/m².a Primary energy demand: 120kWh/m².a
Tigh-Na-Cladach by Gökay Deveci and consultants SPHC	Ten low-energy affordable housing units, one unit Passivhaus-certified – the first in Scotland. No south-facing windows owing to site constraints. DUNOON, ARGYLL, SCOTLAND Treated floor area: 80m² Airtightness: 0.38ach Annual space heat demand: 21kWh/m².a (heat load 10W/m²) Primary energy demand: 99kWh/m².a

(Cont.)

Project	Description	Image
Totnes Passivhaus by Passivhaus Homes Ltd and CTT Sustainable Architect	Adam Dadeby and Erica Aslett's home – retrofit 1970s single family house. TOTNES, DEVON Treated floor area: 162m² Airtightness: 0.2ach Annual space heat demand: 13.3kWh/m².a Primary energy demand: 68kWh/m².a	
Underhill House by Seymour-Smith Architects and consultants SPHC	Single family house in Area of Outstanding Natural Beauty (AONB), as seen on *Grand Designs*. First Passivhaus-certified domestic building in England. COTSWOLDS Treated floor area: 358m² Airtightness: 0.22ach Annual space heat demand: 13kWh/m².a Primary energy demand: 62kWh/m².a	
Viking House by Dudley Marsh Architects with Van Developments and input from Grießbach and Grießbach	Office building for WCR Property Ltd. DOVER Treated floor area: not known Airtightness: 0.3ach Annual space heat demand: 9kWh/m².a Primary energy demand: 103kWh/m².a	
Wimbish Passivhaus Project by Parsons & Whittley	Scheme of 14 social houses for Hastoe Housing Association. WIMBISH, ESSEX. For south-facing units: Treated floor area: not known Airtightness: 0.6ach Annual space heat demand: 12kWh/m².a Primary energy demand: 100kWh/m².a	
Y Foel by JPW Construction. Owner Mark Tiramani	Single private family house. First Passivhaus-certified UK new-build house. WALES Treated floor area: 99.5m² Airtightness: 0.25ach Annual space heat demand: 15kWh/m².a Primary energy demand: 106kWh/m².a	

Glossary of terms*

Terms listed in this glossary are indicated in tinted bold the first time they appear within the text of the book. Tinted bold text in this glossary refers to terms elsewhere in this glossary or to units in the glossary of units.

AECB The Sustainable Building Association, formerly known as the Association for Environment Conscious Building. See Resources.

AECB Silver Standard / AECB Gold Standard Voluntary sustainability standards for buildings developed by the **AECB**. AECB Gold Standard is a Passivhaus-plus energy standard. See www.aecb.net/standards_and_guidance.php.

Air changes per hour (ach) See **Air permeability and air changes per hour (ach)**, below.

Air permeability and air changes per hour (ach) Air permeability is the measure (in $m^3/hr/m^2$) most commonly used in the UK to measure **airtightness**. Passivhaus uses a different measure: air changes per hour. Both conventions are based on a pressure difference of 50 **Pa** (pascals) above and below ambient atmospheric pressure, i.e. under both pressurisation and depressurisation.

$$\text{Air permeability } (q_{50}) = \frac{\text{Flow rate of air entering/exiting building (m}^3\text{/hr)}}{\text{Surface area of thermal envelope (m}^2\text{)}}$$

$$\text{Air changes per hour } (n_{50}) = \frac{\text{Flow rate of air entering/exiting building (m}^3\text{/hr)}}{\text{Ventilation volume, } V_{n50} \text{ (m}^3\text{)}}$$

The ventilation volume, V_{n50}, is the measure of total internal air volume, including volume that is above areas excluded from the **treated floor area (TFA)**. For example, it includes space under staircases, loft space, and spaces used for the **MVHR** unit and hot water store. As such, it needs to be calculated independently of the TFA and should not be confused with V_v, which is simply the ceiling height (assumed by default in the **Passivhaus Planning Package [PHPP]** to be 2.5m) multiplied by the TFA.

Airtightness The degree of leakage of air through the **thermal envelope**. (See **air permeability and air changes per hour**, above). Airtightness is a property of a building.

Annual [specific] primary energy demand The measurement of consumption of primary energy – the energy consumed at source (e.g. a natural gas well) – by a building to meet all its energy needs. This includes all the energy used by every fixed or portable appliance or device within the building. The Passivhaus limit is 120kWh/m².a (kilowatt hours [of primary energy] per square metre [of usable floor area or **treated floor area (TFA)**] per annum. ('Specific' refers to the per-area measurement.) In the **Passivhaus Planning Package (PHPP)**, a ratio, the 'primary energy factor', is used to convert primary energy to delivered energy (energy delivered to the building). For mains electricity this ratio is currently 2.6:1.

Annual [specific] space heat (or cooling) demand The expression used to measure the energy consumed by a building to provide heating to 20°C (or, in hot climates, cooling) of the internal space within the **thermal envelope**. It is measured in kWh/m².a (kilowatt hours per square metre [of **treated floor area**] per annum). ('Specific' refers to the per-area measurement.)

ASHRAE American Society of Heating, Refrigerating and Air-conditioning Engineers. See Resources.

Blown-in A mechanism for 'pumping' in loose insulation through flexible pipes to fully fill a cavity.

BRE Formerly known as the Building Research Establishment, but now simply as BRE. See Resources.

Breathable The ability of a material or building assembly to allow vapour through it by **diffusion**. In practical terms this refers to water vapour, since this is the constituent of air in a building that varies in concentration to any significant degree. A breathable (or breather) membrane is both airtight (it acts as a barrier to air as a whole) and liquid-moisture-tight but **vapour-open**, so will allow water vapour – moisture – through.

BREEAM Building Research Establishment Environmental Assessment Method. A widely recognised environmental assessment method and ratings system mainly used for non-domestic buildings (approx. 200,000 buildings assessed to date). It uses a straightforward scoring system, BREEAM 'Outstanding' being the highest award. There was a domestic new-build equivalent termed 'EcoHomes' launched in 2000, but this expired in April 2012, being replaced by the **Code for Sustainable Homes (CSH)** for new housing.

* Available in pdf format with live links at www.greenbooks.co.uk/passivhaus-handbook

Building assembly A structural part of a building (walls, floor or roof) made up of a number of **building elements**. Referred to as 'Building Element Assembly' in the U-values worksheet of the **Passivhaus Planning Package (PHPP)**.

Building element A single material or object comprising part of the structure of a building, i.e. part of a wall, floor or roof. An 'opaque' building element refers to any building element except windows and doors.

Building Regulations In the UK, these currently (as of 2010) apply in England and Wales and set standards for building construction including energy efficiency. There are 14 technical parts to the regulations (Part A to Part P). Part L relates to the conservation of fuel and power, and Part L1A relates to new dwellings. The edition of Building Regulations previous to 2010 was in 2006, and the next planned update is in 2013. The power to set Building Regulations has now been devolved to Wales, so the 2013 regulations are likely to apply only to England. For copies of Building Regulations see http://www.planningportal.gov.uk/buildingregulations/approveddocuments.

Capillarity (capillary action) Movement of liquid water against gravity, through small pores or capillaries in a material.

Certified Passivhaus Designer An individual who has trained (taken a **Passivhaus-Institut [PHI]**-recognised Certified European Passive House [CEPH] course and PHI CEPH examination) and qualified in the principles and methodology needed to design a Passivhaus. It is also possible to gain Certified Passivhaus Designer status by designing and building a Certified Passivhaus building, although this is generally seen as a harder and potentially riskier way to become a Certified Designer. See www.passivhausplaner.eu/englisch/index_e.html.

Chi-value (χ) Similar to **psi-value (ψ)**, this measures the rate at which heat passes through a material that penetrates another material at a point, where the penetrating material conducts heat better than the surrounding material: for example, a metal bolt, used to mount a balcony, that passes through an external wall. In a Passivhaus chi-value is used to measure heat loss in a point **thermal bridge**. It is measured in **W/K (watts** per **kelvin)**.

Code for Sustainable Homes (CSH) The UK's national standard for the sustainable design and construction of new homes (residential buildings). Dwellings can be rated from Level 3 to Level 6 (L3 to L6), L6 being the most stringent and originally termed '**zero carbon**'. Zero carbon now applies to both L6 and L5, and is currently set to come into force as a statutory requirement in 2016. L3 and L4 are seen as a stepping stone between current UK **Building Regulation** requirements and CSH L5 and L6. L4 is planned to come into force in 2013.

Compact unit A single unit, about the same size and footprint as a large fridge-freezer, that combines **mechanical ventilation with heat recovery (MVHR)**, hot water production/storage and the provision of small amounts of heat, sometimes via supply duct heating (see **supply duct radiator**).

Cooling load Analogous to [specific] **heat load**, this is the power needed to keep a building cool in a hot climate. Whereas [specific] heat load has been defined precisely, in terms of the power per square metre and the temperature difference being maintained (+20°C inside, -10°C outside), no similar quantified definition of cooling load has yet been finalised by the **Passivhaus Institut**.

Desiccant A **hygroscopic** substance, which is used to remove excessive humidity and thus avoid condensation in spaces such as between the panes of a double- or triple-glazed window.

Diffusion The thermal motion – movement driven by temperature – of all liquid and gas particles. The speed of the motion depends on the temperature and the particle size. Diffusion explains how particles move from a place of higher concentration to one of lower concentration (across a **vapour pressure differential** or **gradient**), but it also occurs, more slowly, where there is no pressure differential.

Ducts The pipes that run between the building's **thermal envelope** and the **MVHR** unit and between the MVHR unit and the various supply and extract points within the building. The *intake duct* (sometimes referred to as the 'ambient' duct) takes fresh air from outside into the MVHR. The *supply duct* takes that air (now containing the heat recovered by the MVHR unit) to supply vents in the living room and bedrooms. The *extract duct* takes old air from the bathroom(s) and kitchen back to the MVHR unit. The *exhaust duct* takes the now-cold air back outside. See Chapter 12.

Embodied energy The energy used in the sourcing, manufacture and transport of a material. The way embodied energy is measured varies according to what you choose to include in the calculation – for example, some calculate energy only up to the point the product leaves the factory gate (cradle-to-gate). Measured in **MJ/kg** (megajoules per kilogram).

Energy balance A term to express the simple idea that energy must be input into a building at the same rate that energy is lost from it. Energy losses in buildings take place because heat is conducted out through the building fabric (walls, roofs, windows, etc.) and convected out via gaps in the building fabric, or where it is not recovered via any ventilation system. In order to maintain a temperature difference, enough heat needs to be provided (gained) to compensate for (balance) the losses. Heat is gained (**solar gain**) and lost through the windows. If the windows are designed, installed and oriented optimally, the gains can outweigh the losses. The other source of heat gain is from the building's occupants (**internal heat gains**). Any remaining shortfall has to be provided by a heat source.

Energy returned on energy invested (EROEI) A ratio of the usable energy acquired against the energy that had to be expended to obtain it. We naturally tend to exploit high-EROEI energy sources first. EROEI is sometimes referred to as 'energy return on investment' (EROI).

EnerPHit The **Passivhaus Institut**'s energy performance standard for retrofits. It allows a maximum **annual [specific] space heat demand** of 25kWh/m².a and an upper **airtightness** limit of 1.0ach, if the 0.6ach target can be shown to be impracticable, and also sets requirements for individual elements of a retrofit, should the 25kWh/m².a requirement not be met. Only retrofits in certain climates, including central Europe and the UK, can be certified to the EnerPHit standard.

Fabric Energy Efficiency Standard (FEES) Developed by the Zero Carbon Hub (see Resources), FEES is the first Passivhaus-style fabric energy efficiency standard to be incorporated into the **Code for Sustainable Homes (CSH)** (see Ene 2 on page 40 of the 2010 *Code for Sustainable Homes: Technical Guide*: www.planningportal. gov.uk/uploads/code_for_sustainable_homes_techguide. pdf). Previously, energy performance was defined solely in terms of reductions in carbon emissions. FEES is still under development, and at the time of writing applies only to CSH Levels 5 and 6.

Form factor The ratio of the external area of the **thermal envelope** to the **treated floor area (TFA)**. Form factor is a measure of how compact the build design is. It is broadly similar to the area:volume ratio.

G-value One of the measures of vapour resistance: see **vapour permeability**. G-value is not to be confused with g-value (see below).

g-value A measure of the percentage of energy from the sun that passes through the glazed unit to reach the interior. In the USA, g-value is known as the solar heat gain coefficient (glazing), or SHGC-glazing.

Ground source heat exchanger (GSHX) Used with **mechanical ventilation with heat recovery (MVHR)** to preheat and pre-cool air via pipes in the ground, where temperatures are more constant than above-ground.

[Specific] Heat load The peak power, in W/m² (**watts per square metre**) of **treated floor area (TFA)**, needed to maintain 20°C inside when the outside temperature is -10°C. ('Specific' refers to the per-area measurement.) The **Passivhaus Planning Package (PHPP)** calculates this at -10°C with overcast conditions and in sunny weather, and takes the more pessimistic of the two scenarios.

Heat main A system of insulated pipes that run between buildings, enabling the use of a large-scale heat source that would be too big for a single building. The heat main transports the heat from the point of generation to the point of use. Such infrastructure is common in Denmark and Holland, where heat from electricity power plants runs on natural gas or biomass; heat is thereby used efficiently instead of being wasted as it is in the UK.

Home Energy Rating System (HERS) A measure of (primarily) energy efficiency; mainly used in the USA. A HERS index of 100 means a home meets the code standard that is based on a standard US house; a HERS index of 70 means the home is 30-per-cent better than the code standard. The report generated advises on potential improvements to an existing property. A home-energy assessor visits the building, and tests carried out normally include a site blower test (pictured on page 125).

Hygroscopicity The property of a material to absorb, retain and release moisture from the ambient air. Materials that readily do this are often described as hygroscopic.

Indoor air quality (IAQ) The quality of air within buildings, in regard to both health and comfort. Indoor air often contains a complex mixture of contaminants and common pollutants, including smoke, volatile organic compounds (VOCs) and moulds. The level of carbon dioxide (CO_2) in indoor air also relates to IAQ, and is an accepted marker for the wider mix of potential indoor air pollutants. **ASHRAE** issues guidelines on acceptable IAQ.

Intelligent membrane A membrane with humidity-variable characteristics, meaning that it is more **vapour-open** when there is a higher average **relative humidity (RH)** on the inside and outside of the membrane, thus allowing more drying out of the building fabric.

Internal heat gains The heat gains in a building from its occupants and the use of appliances within the **thermal envelope**. See **energy balance**.

Interstitial condensation Condensation that occurs within a **building assembly**, when warm moist air (generally from inside a heated building in winter) penetrates into the assembly, meets a cold surface and condenses. This can cause serious structural damage and affect insulation performance, especially if the condensation is not able to dry out.

kelvin (K)   See **K** in glossary of units.

Lambda value (λ), also known as k-value   A measure of **thermal conductivity**, measured in **W/mK** (watts per metre [depth] per degree **kelvin**). The inverse measurement is **thermal resistivity**. See also Appendix B. Lambda 90/90 ($\lambda_{90/90}$) values are thermal conductivity values that have been calculated according to the Lambda 90/90 convention, which means that 90 per cent of the test values show a lower conductivity than the stated value, to a statistical confidence level of 90 per cent. These are the values adopted in the UK for Passivhaus calculations. Lambda 90/90 values refer to materials that are factory-produced and regularly tested. Materials that are **blown-in** on-site need density checks, as conductivity is strongly density-dependent. It is much harder to obtain a 90/90 value for materials that are made on-site, such as hemp and lime.

Leadership in Energy and Environmental Design (LEED)   A US sustainability rating system, broad-based and internationally recognised. See www.usgbc.org.

Mechanical ventilation with heat recovery (MVHR), also known as heat recovery ventilation (HRV) or comfort ventilation   A whole-house ventilation system that takes out heat from the old (exhaust) air and gives it to the new (intake) air. Fresh air is delivered to living areas (e.g. living room and bedrooms) and extracted from kitchens and bathrooms. MVHR units do not supply new heat into the supplied air. However, a **supply duct radiator** can be used to add heat to the new air after it leaves the MVHR unit. See Chapter 12.

n_{50}   The term for describing the **air changes per hour (ach)** at a pressure of 50**Pa** (pascals) above and below ambient atmospheric pressure, applied to the inside of a building in an **airtightness** test.

Net energy   The remaining energy available to society after the energy needed to obtain it has been subtracted. This is sometimes expressed as a ratio and is referred to as **energy returned on energy invested (EROEI)**. 'Peak net energy' (a more accurate term than the more commonly used 'peak oil') refers to the maximum rate at which energy can be extracted from a source; that rate being constrained by physical rather than economic factors. See http://netenergy.theoildrum.com.

Off-gassing   The evaporation of volatile chemicals (including volatile organic compounds [VOCs]) at atmospheric pressure and room temperature. Off-gassing of potentially harmful chemicals occurs in many modern, mainstream building materials, such as paints, varnishes and chipboards. Concerns are sometimes expressed about off-gassing from some insulation materials. Off-gassing becomes a problem in enclosed environments without sufficient air changes.

Parging   A term (in a Passivhaus context) for plastering of rough walls, etc., to seal for **airtightness**.

Parts per million (ppm)   Used to measure atmospheric concentrations of pollutants and other gases, including carbon dioxide (CO_2).

Passive stack ventilation   A means of ventilating without using mechanical fans. Utilises natural 'stack' effects from the temperature difference between the inside and outside of the building and from wind passing over the building (creating suction). Systems use physical vertical pipes to provide exit points at a high level.

Passivhaus Institut (PHI)   The independent foundation established in Germany in 1996 to develop, promote and protect the Passivhaus standard. Known as the Passive House Institute in English-speaking countries. See Resources.

Passivhaus methodology   Using the knowledge contained in the Certified European Passive House (CEPH) Passivhaus Designer course, including modelling in the **Passivhaus Planning Package (PHPP)**, to design a building that performs to a defined ultra-low-energy standard.

Passivhaus Planning Package (PHPP)   The energy-modelling design tool created by the **Passivhaus Institut (PHI)** to accurately predict energy performance. It is the basis for designing and certifying Passivhaus and **EnerPHit** builds. See Chapter 7.

Peak net energy   See **net energy**.

Psi-value (ψ)   Similar to **chi-value (χ)**; a measure of the rate at which energy passes through a length of material. In a Passivhaus it is used to measure heat loss in a linear **thermal bridge**. It is measured in **W/mK** (watts per [lateral] metre per degree **kelvin**). The psi-value of specific **building elements** or components is referred to thus:
- ψ_{spacer}: psi-value of the **spacer** between panes in a double- or triple-glazed window.
- $\psi_{installation}$: psi-value of the junction between a window frame and the wall.

Different conventions for measuring the psi-value are described thus:
- ψ_i: internal dimensions of the junction between building elements (general convention)
- ψ_e: external dimensions of the junction between building elements (Passivhaus convention)

See also **U-value**.

q₅₀ The term for describing the **air permeability**, in m³/hr/m², at a pressure of 50Pa (pascals) above and below ambient atmospheric pressure, applied to the inside of a building in an **airtightness** test.

Regulated (carbon) emissions A concept in the **Code for Sustainable Homes (CSH)**. Refers to emissions resulting from energy use to provide only heating, hot water, fixed lighting, pumps and fans. See also **unregulated (carbon) emissions**.

Relative humidity (RH) A measure of the quantity of water vapour in a given volume of air, expressed as a percentage of the maximum quantity of water vapour that volume can contain before it becomes saturated (after which condensation will occur).

Solar gain The amount of energy from the sun captured through glazing.

Spacer The dividing strip along the edge of a double- or triple-glazed unit that separates each pane. Warm-edge spacers are made from material or materials with a lower conductivity. The energy performance of a spacer is measured by its **psi-value** (ψ).

Specific heat capacity The amount of heat required to change the temperature of a unit of a material by a given amount. (In standard metric units, it is the number of joules required to raise 1 gram of the material by 1 degree **kelvin**.) Specific heat capacity is a measure of a material's **thermal mass**.

Specific heat load See **[Specific] Heat load**.

Standard Assessment Procedure (SAP) The UK government's recommended tool for determining the energy rating of dwellings (residential buildings), first published in 1995 and subsequently updated. It evolved from the National Home Energy Rating (NHER) scheme and was devised by **BRE**.

Summer bypass A control on the **MVHR** unit to bypass the heat recovery function so you can continue to use it to ventilate at warmer ambient temperatures.

Supply duct radiator, also known as in-line duct radiator A small radiator inserted into the supply duct just after it leaves the **MVHR** unit. It adds a small amount of heat into the ventilation system's supply duct. See **ducts**.

Thermal bridge Commonly known as a cold bridge. A gap in insulation that allows heat to 'short-circuit' or bypass it. This occurs when a material with relatively high conductivity interrupts or penetrates the insulation layer. Thermal bridges exist in point and linear form. See Chapter 8.

Thermal bypass A type of **thermal bridge** caused by air movement in the insulation layer, for example in an unfilled cavity in a typical UK cavity wall or in poorly laid loft insulation, leading to the transfer of heat.

Thermal comfort Defined by Dr P. Ole Fanger as "the condition of mind which expresses satisfaction with the thermal environment". Dr Fanger identified thermal comfort as being determined by: air temperature, 'radiant' temperatures (the temperatures of walls, floor and ceiling), air movement (draughts), temperature stratification (differences in temperature from floor to ceiling), **relative humidity (RH)**, the insulative value of clothing ('Clo Value') and physical activity level ('Met Value'). These form the basis of **ASHRAE**'s standards for thermal comfort. In contrast, 'adaptive models' describe thermal comfort as a function of physiological, psychological and behavioural factors.

Thermal conductivity, also known as conductivity A material's ability to transmit heat, measured by the **lambda value** (λ). Unlike **U-value**, the lambda value of a material remains the same irrespective of the thickness of the material. Lambda values do sometimes vary with temperature. See Appendix B for typical values for common building materials.

Thermal envelope The area of floors, walls, windows and roof or ceiling that contains the building's internal warm/heated volume.

Thermal mass The ability of a body of material to absorb, store and subsequently release heat (due to its **specific heat capacity** and its mass).

Thermal resistivity, also known as resistivity (but not to be confused with vapour resistivity – see **vapour permeability**) A material's ability to resist the passage of heat. It is the mathematical inverse of **thermal conductivity**, and is measured in **Km/W** (**kelvin** metres per **watt**). Two resistivity values need to be entered for the calculation of **U-values** in the PHPP:

- R_{si} describes the resistivity of the static air* on the interior surface of the material.
- R_{se} describes the resistivity of the static air* on the exterior surface of the material.

(*The first millimetre of air on a surface remains static even in windy conditions and therefore has insulation properties.)

Thermal store Provides the same function as a conventional hot water tank or store, but the hot water is never directly used in the taps. Instead, cold mains water is fed through a coil in the thermal store and is heated by the surrounding water. Thermal stores can work at a lower operating temperature than conventional tanks, and can also be connected to 'uncontrolled' heat sources, such as the back boiler of a wood burner. See Appendix A.

Transfer path   A 20mm gap under a door, or a hidden 10mm gap cut into the top of the architrave and door frame, to allow air to move between supply or extract rooms and the common interconnecting spaces (in a domestic house, usually the stairwell and hallways).

Treated floor area (TFA)   A convention for measuring usable internal floor area within the **thermal envelope** of a building. See Chapter 7, page 94.

U-value   A measure of the ease with which a material or **building assembly** allows heat to pass through it; in other words, how good an insulator it is. The lower the U-value, the better the insulator. The U-value is used to measure how much heat loss there is in a wall, roof, floor or window, and is measured in **W/m²K** (**watts** per square metre per degree **kelvin**). When referring to window U-values, the following convention is used:
- U_f – the U-value of the window frame.
- U_g – the U-value of the glazing.
- U_w – the U-value of the whole window. This value is often quoted by window manufacturers for a standard size and configuration of window (1230mm high by 1480mm wide, with a centre mullion, one fixed and one opening casement) and depends on U_f, U_g and the **spacer psi-value**.
- $U_{w, installed}$ – similar to U_w but also takes into account the losses as a result of the **thermal bridge** between the window and the wall, which is quantified by the installation psi-value. $U_{w, installed}$ is calculated by the **Passivhaus Planning Package (PHPP)**.

Ultra-low-energy   This is not a formal standard, but a term we have chosen to use in this book to refer to buildings that require up to 40kWh/m².a (kilowatt hours per square metre [of **treated floor area**] per annum) for space heating and have an **airtightness** of 1.5 **air changes per hour (ach)** at 50**Pa** (pascals) above and below ambient atmospheric pressure. The **AECB Silver Standard** sets a space-heating limit of 40kWh/m².a and **zero carbon** sets a limit of around 50kWh/m².a in Passivhaus terms (i.e. adjusted for the difference in convention used for measurement of floor area).

Unregulated (carbon) emissions   A concept in the **Code for Sustainable Homes (CSH)**, referring to emissions resulting from energy use to provide heating, hot water, fixed lighting, pumps and fans, as well as from energy use for cooking and other household electrical appliances. In Passivhaus, the energy equivalent of this concept is **annual [specific] primary energy demand**. See also **regulated (carbon) emissions**.

Vapour barrier, also known as vapour-closed   A material that is near-impermeable to water vapour, e.g. aluminium foil.

Vapour-closed   See **vapour barrier**.

Vapour-open   A material that is permeable to water vapour, e.g. a 'breathable' or 'breather' membrane (these are both airtight and liquid-moisture-tight but vapour-open).

Vapour permeability   The degree to which a material facilitates the passage of water vapour through it, measured by four different values: vapour resistivity (r-value; units: **MNs/gm** – meganewton seconds per gram metre); vapour resistance (G-value; units: **MNs/g** – meganewton seconds per gram); water vapour resistance factor (μ-value; no units); equivalent air layer thickness (Sd-value; units: m – metres).

Vapour pressure differential/gradient   The pressure exerted by a vapour, or gas (for example, water vapour) relates to its temperature and the concentration of the particles. Where there is a higher vapour pressure on one side of a material than on the other, the difference between the two is the vapour pressure differential. Vapour will move by **diffusion** across a **vapour-permeable** membrane, driven by the pressure difference. So water vapour will move from the side with higher humidity (high vapour pressure) to the side with lower humidity (low vapour pressure).

Watt (W)   See **W** in glossary of units.

Window schedule   A list of all the windows in a building; usually produced by an architect to communicate to other parties in the build project what windows are required.

Zero carbon – a UK energy target for housing, currently set for enforcement in 2016. It currently applies to both Levels 5 and 6 (L5 and L6) of the **Code for Sustainable Homes (CSH)**. In broad terms, it can be considered to mean that there are no net annual greenhouse gas emissions resulting from energy use in a dwelling. Any emissions created are offset by those 'saved' using on-site (or possibly communal if the dwelling is part of an estate) renewable capacity, which feeds electricity back to the grid. It now includes the **Fabric Energy Efficiency Standard (FEES)**.

Glossary of units

All units used in the book are included in this glossary, and are tinted bold the first time they appear within the text of the book. Tinted bold text in this glossary refers to terms elsewhere in this glossary or to terms in the glossary of terms.

K   Kelvin, a unit of measurement for temperature. One degree kelvin = one degree Celsius, but zero degrees kelvin (0K) is absolute zero. 0°C is approx. 273K.

Km/W   Kelvin metres per watt. The unit of thermal resistivity.

kWh   Kilowatt hour, a unit of energy. For example, a one-kilowatt electric fire left on for one hour would use 1kWh (1,000**W**) of energy. 1kWh = 3.6MJ (megajoules).

kWh/m².a   Kilowatt hours per square metre per annum. Measures energy used annually per square metre of usable or treated floor area (TFA). This is one of the key units of measure in Passivhaus. By defining energy use in terms of each square metre of floor area, it allows us to make a meaningful comparison of the energy use of buildings of different sizes. See annual [specific] space heat demand.

kWp   Kilowatt peak. A measure of the maximum power output of photovoltaic (solar) panels.

m³/hr   Cubic metres [of air] per hour. The unit of measure used to describe ventilation rates. Passivhaus ventilation calculations are based on providing 30m³ per person per hour. This rate is necessary to keep carbon dioxide (CO_2) levels well below 1000 parts per million (ppm).

m³/hr/m²   Cubic metres [of air] per hour per square metre [of thermal envelope area]. The unit of measure of air permeability.

MJ/kg   Megajoules per kilogram. Unit of measure of embodied energy.

MJ/m²   Megajoules per square metre. A megajoule (MJ) is a unit of energy, equivalent to 1,000,000J (joules). 3.6MJ = 1**kWh**.

MNs/g   Meganewton seconds per gram. The unit of measure of vapour resistance (G-value). See vapour permeability.

MNs/gm   Meganewton seconds per gram metre. The unit of measure of vapour resistivity (r-value). See vapour permeability.

Pa   Pascal, the SI (International System of Units) measurement of force per unit area. 1Pa is 1 newton per square metre.

W   Watt, a unit of power. The rate at which energy transfers (is 'consumed' or 'generated'). For example, a typical low-energy lightbulb uses 15W; an electric kettle 3kW (kilowatts, i.e. 1,000W). 1W = 1 joule per second.

W/K   Watts per degree kelvin [temperature difference between inside and outside the thermal envelope]. Used to quantify chi-value (χ) in point thermal bridges.

Wh/K per m² (as cited in the PHPP)   Watt hours per kelvin per square metre [of treated floor area (TFA)]. 1 watt hour (Wh) is one-thousandth of a **kWh** (kilowatt hour): for example, a 1W LED light bulb left on for 1 hour would use 1Wh of electricity.

Wh/m³   Watt hours per cubic metre [of air moved]. Used by the Passivhaus Institut (PHI) to measure the electrical efficiency of MVHR units. 1Wh (watt hour) is one-thousandth of a kWh.

W/m²   Watts per square metre [of treated floor area (TFA)]. Used to measure heat load.

W/m²K   Watts per square metre [of the material/assembly in question] per degree kelvin [temperature difference between inside and outside the thermal envelope]. Used to quantify U-value.

W/mK   Watts per metre per degree kelvin. For linear thermal bridges, this is watts per metre length of the thermal bridge per degree kelvin temperature difference. (Used to measure psi-value [ψ].) For thermal conductivity this is watts per metre thickness/depth of material per degree kelvin temperature difference. (Used to measure lambda value [λ].) In both cases the temperature difference measured is that between inside and outside the thermal envelope.

Notes*

Introduction

1. Research by Leeds Metropolitan University and other evidence shows the performance gap between 'as designed' and 'as constructed' can be very large – sometimes double. See Zero Carbon Hub (2010). 'Carbon compliance for tomorrow's new homes: A review of the modelling tool and assumptions, Topic 4 – Closing the gap between design and built performance', www.zerocarbonhub.org/resourcefiles/TOPIC4_PINK_5August.pdf.

Chapter 1

1. Assumes (1) that the floor area of the average UK home is approx. 95m^2, which is roughly equivalent to 75m^2 when measured according to the definition of treated floor area (TFA) used by Passivhaus; and (2) that space heating consumes an average 15,000kWh per annum.
2. Assumes (1) that the floor area of the average UK home is approx. 95m^2, which is roughly equivalent to 75m^2 when measured according to the definition of TFA used by Passivhaus; and (2) that an average 20,500kWh per annum of delivered energy is used for heating and hot water, and 3,500kWh per annum of delivered energy for all other uses. If natural gas is used for heating and hot water, and electricity for all other uses, ratios of 1:1 (gas) and 1:2.6 (electricity) respectively are used to convert from delivered to primary energy.
3. The Passivhaus Institut's EnerPHit standard is set out in full on the PHI website: www.passiv.de/en/03_certification/02_certification_buildings/04_enerphit/04_enerphit.htm.
4. Department for Communities and Local Government (2010). *Code for Sustainable Homes: Technical Guide*, Ene 2, p.40, www.planningportal.gov.uk/uploads/code_for_sustainable_homes_techguide.pdf.

Chapter 2

1. See: Heinberg, R. (2005). *The Party's Over*. Second edition. Clairview Books: Forest Row; Greer, J. M. (2008). *The Long Descent*. New Society Publishers: Gabriola Island, BC, Canada; Hirsch, R. L., Bezdek, R. and Wendling, R. (2005). 'Peaking of world oil production: Impacts, mitigation, and risk management' ('the Hirsch Report'). Prepared for the US Department of Energy (DoE) National Energy Technology Laboratory (NETL) by Science Applications International Corporation (SAIC), www.netl.doe.gov/publications/others/pdf/Oil_Peaking_NETL.pdf.
2. See Stern, D. (2010). 'The role of energy in economic growth'. Working paper, Centre for Climate, Economics & Policy (CCEP), Crawford School of Economics and Government, The Australian National University, Canberra, http://ccep.anu.edu.au/data/2010/pdf/wpaper/CCEP-3-10.pdf. Overview at The Oil Drum: www.theoildrum.com/node/8476.
3. International Passive House Association (iPHA) / Passivhaus Institut (PHI) (2010). 'Active for more comfort: The Passive House' ['The Passive House Brochure'], p.39 (pdf), p.36 (hard copy), www.passivehouse-international.org/index.php?page_id=70.
4. The Passive-On Project CD (2007), www.passive-on.org/en/cd.php.
5. International Passive House Association (iPHA) / Passivhaus Institut (PHI) (2010). 'Active for more comfort: The Passive House', p.37 (pdf), p.34 (hard copy), www.passivehouse-international.org/index.php?page_id=70.
6. Newman, N., bere:architects (2011). 'Case study evidence for reduced whole life costs . . . in various price scenarios'. UK Passivhaus Conference, 25 October, http://ukpassivhausconference.org.uk/sites/default/files/Nick%20Newman.pdf. Also see the Passivhaus Cost Project: www.phcp.org.uk.

Chapter 3

1. Information about this is available on the International Passive House Association (iPHA) website: www.passivehouse-international.org/index.php?group=1&level1_id=76&page_id=246&lang=de.
2. See the FAQ about PHIUS+ project certification at: www.passivehouse.us/passiveHouse/PHIUSPlusDocs/PHIUS+FAQ.pdf.
3. See www.passiv.de/en/03_certification/01_certification_components/01_component_database.htm for a current list of certified components.
4. See link in note 3 above for a current list of certified components.
5. Courses within Europe are organised by Certified European Passive House (CEPH). See

* Available in pdf format with live links at www.greenbooks.co.uk/passivhaus-handbook

www.passivhausplaner.eu/englisch/planer_werden.html and http://eu.passivehousedesigner.de. See also www.passivehouse-international.org/index.php?group=1&level1_id=194&page_id=221&lang=de.
6 WARM: Low Energy Building Practice: www.peterwarm.co.uk/certify.

Chapter 4

1. Collinson, P. (2011). 'Self-build: it's time to go Dutch'. *Guardian*, 25 November, www.guardian.co.uk/money/2011/nov/25/self-build-go-dutch.
2. Goodier, C. I. and Pan, W. (2010). 'The future of UK housebuilding'. RICS: London.

Chapter 5

1. Hammond, J. and Jones, C. (2011). 'Inventory of Carbon and Energy (ICE)'. Sustainable Energy Research Team (SERT), Department of Mechanical Engineering, University of Bath, UK. Copy available at http://perigordvacance.typepad.com/files/inventoryofcarbonandenergy.pdf. This document is added to and updated, but as such is available only to those with University of Bath log-in at http://wiki.bath.ac.uk/display/ICE.
2. Sustainable Homes (1999). 'Embodied energy in residential property development: A guide for registered social landlords', p.3, www.sustainablehomes.co.uk/Portals/63188/docs/Embodied%20Energy.pdf.
3. For more information see Lifetime Homes: www.lifetimehomes.org.uk.
4. See Grant, N. and Clarke, A. (2010). 'Biomass – a burning issue', AECB discussion paper, www.aecb.net/PDFs/Biomass_A_Burning_Issue_September_2010.pdf.
5. Zero Carbon Hub (2011). 'Carbon compliance: Setting an appropriate limit for zero carbon new homes', www.zerocarbonhub.org/resourcefiles/CC_TG_Report_Feb_2011.pdf.
6. Department for Communities and Local Government (2010). *Code for Sustainable Homes: Technical Guide*, Ene 2, p.40, www.planningportal.gov.uk/uploads/code_for_sustainable_homes_techguide.pdf.
7. See the helpful summary at www.zerocarbonhub.org/resourcefiles//Dwelling_Classifications.pdf.
8. Zero Carbon Hub Ventilation and Indoor Air Quality (VIAQ) Task Group (2012). 'Mechanical ventilation with heat recovery in new homes: Interim report', www.zerocarbonhub.org/resourcefiles/ViaqReport_web.pdf.
9. Passivhaus Trust (2011). 'Passivhaus and zero carbon: Technical briefing document', Appendix A, p.7, www.passivhaustrust.org.uk/UserFiles/File/Technical%20Papers/110705%20Final%20PH%20ZC%20Brief.pdf.
10. Carbon Trust (2011). 'Conversion factors: Energy and carbon conversions – 2011 update', www.carbontrust.co.uk/cut-carbon-reduce-costs/calculate/carbon-footprinting/pages/conversion-factors.aspx. See also http://archive.defra.gov.uk/environment/business/reporting/pdf/110707-guidelines-ghg-conversion-factors.pdf.
11. See Dadeby, A. (2012). 'Energy returned on energy invested', in Raymond De Young, R. and Princen, T. (eds.) *The Localization Reader: Adapting to the coming downshift*. The MIT Press: Cambridge, MA.
12. Hall, C. A. S., Tharakan, P. and Ko, J. Y. (2002). 'A review of global use and availability of energy sources'. Third biennial international workshop, Advances in energy studies: Reconsidering the importance of energy, Porto Venere, Italy, 24-8 September. (For background information, see www.chim.unisi.it/chimica/portovenere.html.)
13. Tainter, J. (1990). *The Collapse of Complex Societies*. Paperback edition. Cambridge University Press: Cambridge.
14. The Oil Drum is a good resource for recent research on EROEI. See www.theoildrum.com/tag/eroi.
15. Homer-Dixon, T. (2007). *The Upside of Down: Catastrophe, creativity and the renewal of civilisation*. Souvenir Press: London.
16. For more information on TEQs see www.teqs.net.

Chapter 6

1. The Passivhaus extension project pictured on page 82 will be featured on our related website: www.passivhaushandbook.com.
2. Fleming, D. (2010). *Lean Logic: A dictionary for the future and how to survive it*. Privately published. Available from www.leanlogic.net.
3. McCloud, K. (2006). *Grand Designs Handbook: The blueprint for building your dream home*, Collins: London, pp.34-5.
4. The Passivhaus Institut's user handbooks can be downloaded from the PHI website: www.passiv.de/de/05_service/03_fachliteratur/030300_nutzerhandbuch/030300_nutzerhandbuch.htm.

Chapter 7

1. Passivhaus Institut (PHI) (2012). 'Certification criteria for certified Passive House glazings and transparent components', Version 2.0 E, http://passiv.de/downloads/03_certification_criteria_transparent_components_en.pdf.
2. Ibid.

Chapter 8
1. Siddall, M. (2009). 'Thermal bypass: The impact of natural and forced convection upon building performance', *Green Building Magazine*, Summer 2009, Volume 19, No. 1, , www.aecb.net/PDFs/Impact_of_thermal_bypass.pdf.
2. See http://windows.lbl.gov/software/therm/therm.html.
3. These are currently offered by the AECB as part of its CarbonLite Programme. See Resources.

Chapter 9
1. Fraunhofer Institute for Building Physics (IBP), Stuttgart. Source: DBZ [Bund Deutscher Zimmermeister (Association of German Master Carpenters)] 12/89, p.1639ff.
2. Limb, M. J. (2001). 'A review of international ventilation, airtightness, thermal insulation and indoor air quality criteria', Technical Note 55. Air Infiltration and Ventilation Centre (AIVC): Coventry, UK, www.aivc.org/frameset/frameset.html?../Publications/Technical_reports/TN55.htm~mainFrame.
3. Langmans, J., Klein, R. and Roels, S. (2010). 'Air permeability requirements for air barrier materials in Passive Houses'. Paper presented at the Fifth International BUILDAIR Symposium, 'Building and Ductwork Airtightness', 21–22 October, Copenhagen, Denmark.

Chapter 10
1. For more on monolithic membranes and roof constructions, see: www.ecologicalbuildingsystems.com/workspace/downloads/proclimaRoofRenovationStudy2011.pdf
2. Sterling, E. M., Arundel, A. and Sterling, T. D. (1985). 'Criteria for human exposure to humidity in occupied buildings'. ASHRAE Transactions, 1985, Vol. 91, Part 1, www.sterlingiaq.com/photos/1044922973.pdf. Also published in the *2000 ASHRAE Handbook: HVAC Systems and Equipment*.
3. Ibid.

Chapter 11
1. Passivhaus Institut (PHI) (2012). 'Certification criteria for certified Passive House glazings and transparent components', Version 2.0 E, http://passiv.de/downloads/03_certification_criteria_transparent_components_en.pdf.

Chapter 12
1. Passivhaus Institut (PHI) 2001/2. CEPHEUS: 'Measurements and evaluation of Passive House apartment buildings in the Marbachshöhe neighbourhood of Kassel, Germany'. (Only available in German from the PHI. CEPHEUS – Cost-Efficient Passive Houses as European Standards – assisted this multi-storey tenement project.)
2. Mahdavi, A. and Doppelbauer, E-M. (2010). 'A performance comparison of passive and low-energy buildings', in *Energy and Buildings* 42, 1314-19, http://iristor.vub.ac.be/patio/ARCH/pub/fdescamp/bruface/literature/2012/nzeb/PerformanceComparison.pdf.
3. Paul Jennings of Air Leakage, Detailing and Awareness Services (ALDAS).

Chapter 13
1. This took place on 9-11 September 2011 and involved 19 households in and around Totnes, Devon, that had built or retrofitted their homes to improve their sustainability.
2. See www.passivhausblog.co.uk.
3. Lynch, H. (2012). 'Passivhaus in the UK: The challenges of an emerging market – a case study of innovation using mixed-method research'. PhD dissertation, University College London.

Chapter 14
1. Cullingworth, B. and Nadin, V. (2006). *Town and Country Planning in the UK*. Fourteenth edition. Routledge: London.
2. See www.aecb.net/PDFs/carbonlite/AECB_VOL3_EnergyStandard_V6FINAL.pdf.
3. Directive 2002/91/EC on the Energy Performance of Buildings: http://eur-lex.europa.eu/LexUriServ/LexUriServ.do?uri=OJ:L:2003:001:0065:0071:EN:PDF.
4. Cahill, K. (2001). *Who Owns Britain*. Canongate Books: London.
5. See Zero Carbon Homes Task Force: www.theyworkforyou.com/wrans/?id=2011-03-29a.49878.h. Also Home Builders Federation (HBF) Zero Carbon and Sustainability: www.hbf.co.uk/policy-activities/government-policy/zero-carbon-sustainability/. The HBF represents the interests of the large home developers.

Resources*

This section provides current useful resources. However, Passivhaus is a fast-moving area, so we have also set up a website, **www.passivhaushandbook.com**, which we aim to keep updated with links to useful new resources.

Organisations

AECB
The Sustainable Building Association, formerly known as the Association for Environment Conscious Building. A network of individuals and organisations with a common aim of promoting sustainable building. For members there are useful discussion and technical forums, information on the AECB Gold Standard (with useful construction details), the opportunity to attend the annual conference, and discounts on CarbonLite courses.
0845 4569 773
www.aecb.net
www.aecb.net/carbonlite/phpp.php to purchase the Passivhaus Planning Package (PHPP) software
www.carbonlite.org.uk/carbonlite/courses.php for AECB CarbonLite courses

ASHRAE
American Society of Heating, Refrigerating and Air-conditioning Engineers. An international organisation whose aim is to increase knowledge about heating, ventilation, air conditioning and refrigeration.
www.ashrae.org

BRE
Formerly known as the Building Research Establishment, but now known simply as BRE. Conducts research and offers consultancy for the built environment sector, and accreditation for Code for Sustainable Homes (CSH) and BREAM sustainability standards.
01923 664000
www.bre.co.uk
www.bre.co.uk/training.jsp for training and accreditation
www.passivhaus.org.uk for specific services on Passivhaus, including UK Passivhaus climate data:
www.passivhaus.org.uk/page.jsp?id=38

Centre for Alternative Technology (CAT)
Expertise in natural materials and sustainable building, including small-scale renewable energy generation. Vocational and postgraduate courses available.
01654 705950
www.cat.org.uk
CAT Graduate School: 01654 704985; http://gse.cat.org.uk
Information centre: 01654 705989; http://info.cat.org.uk

International Passive House Association (iPHA)
The global Passivhaus network to promote the standard, increase public understanding and exchange knowledge. Members of the Passivhaus Trust in the UK automatically become members of the iPHA. The iPHA coordinates the Passive House Open Days internationally.
www.passivehouse-international.org

Passipedia
Run by the iPHA, bringing together Passivhaus-relevant articles from around the world on a Wiki-style site. Members of iPHA get access to more in-depth information. PHPP software is available for purchase.
www.passipedia.org

Passive House Association Ireland (PHAI)
Aims to promote, educate, facilitate and develop a strong identity, understanding and demand for the Passive House concept.
www.phai.ie

Passive House Institute US (PHIUS)
Provides training, education and research for all US climate zones. PHIUS is not currently accredited by the Passivhaus Institut, but has its own certification scheme.
www.passivehouse.us

Passivhaus Institut (PHI), sometimes referred to as the Passive House Institute
Information and research, including publications (although largely in German). The 17th International Passive House Conference will be held on 19-20 April 2013 in Frankfurt, and includes an exhibition of the latest Passivhaus-suitable products. See www.passivhaustagung.de/siebzehnte/Englisch/index_eng.html
http://passiv.de/en/index.php (PHI home page in English)

* Available in pdf format with live links at www.greenbooks.co.uk/passivhaus-handbook

www.passivhausplaner.eu/englisch/planersuche.php for a global directory of qualified Passivhaus Designers and Consultants

www.passivehouse-trades.org for information on the new Certified Passivhaus Tradesperson qualification. Currently there are several certified tradespeople in Ireland but none in England, Wales or Scotland

www.passiv.de/en/03_certification/03_certification.htm for a database of certified building components, buildings, and Passivhaus Designers, Consultants and tradespeople

Passivhaus Trust

A subsidiary company of the AECB, this is a non-profit-making organisation for promoting Passivhaus and protecting the Passivhaus standard in the UK. UK Passivhaus Awards were run for the first time in 2012. Organises the UK Passivhaus Annual Conference.
020 7704 3502
www.passivhaustrust.org.uk
www.ukpassivhausconference.org.uk for information on the Annual Conference

Royal Institute of British Architects (RIBA)

Professional association of architects in the UK. The RIBA shop sells standard appointment contracts and stocks various advisory leaflets.
020 7580 5533
www.architecture.com

Zero Carbon Hub

Facilitates the delivery of zero-carbon and low-carbon homes in the UK. Offers useful publications and conducts research on ultra-low-energy housing.
0845 888 7620
www.zerocarbonhub.org

Publications

IBO (The Austrian Institute for Building and Ecology Ltd) (2009). 'IBO Book': *Passivhaus-Bauteilkatalog: Ökologisch bewertete Konstruktionen / Details for Passive Houses: A catalogue of ecologically rated constructions* (German and English edition), Springer: New York.
Extremely useful, if expensive, reference book of construction details, although not always well translated and some materials are unfamiliar to the UK. Many details include a psi-value (thermal bridge) calculation.

Cullingworth, B. and Nadin, V. (2006). *Town and Country Planning in the UK*. Fourteenth edition. Routledge: Abingdon.
Detailed and comprehensive description of the UK's planning system.

James, M. (2010). *Recreating the American Home: The Passive House approach*. Low Carbon Productions: Larkspur, California.
An account of ten inspirational Passivhaus projects across the USA

Klingenberg, K., Kernagis, M. and James, M. (2008). *Homes for a Changing Climate: Passive Houses in the US*. Low Carbon Productions: Larkspur, California.
An overview of nine pioneering US projects in a range of climates.

Passivhaus Institut (PHI) (2010). *The Passive House Architecture Award Book*. Passivhaus Institut: Darmstadt. Available from the iPHA website (see page 249)
Summary project details and photographs of 24 projects around the world.

Royal Institute of British Architects (2009). *A Client's Guide to Engaging an Architect: Guidance on hiring an architect for your project*. RIBA Publishing: London.

Speer, R. and Dade, M. (2009). *How to Get Planning Permission*. Fourth edition. Ovolo Books: Kimbolton. A helpful, accessible guide to help self-builders get planning permission.

Passivhaus Designer/Consultant training courses

For a full list in Europe, visit **www.passivhausplaner.eu/englisch/phkurs/kurssuche.php**. In the UK there are three main approved training courses:
- AECB CarbonLite Passivhaus Designer course, held at a variety of locations. See www.carbonlite.org.uk/carbonlite/course_detail.php?cId=15
- BRE (Watford). See www.bre.co.uk/eventdetails.jsp?id=4176
- Strathclyde University (Glasgow) in cooperation with Scottish Passive House Centre (SPHC). See **www.sphc.co.uk/certified-european-passive-house-designer**

Passivhaus Certifiers

The following are PHI-accredited Passivhaus Certifiers in the UK and Ireland:
- BRE. See **www.bre.co.uk/accreditation/page.jsp?id=2648** (Note: this is a BRE certification scheme and relies heavily on Passivhaus Designer/Consultant support)
- Cocreate. See **www.cocreateconsulting.com/services/passivhaus**

- MosArt Ltd. See www.mosart.ie
- WARM: Low Energy Building Practice. See www.peterwarm.co.uk/certify

For current details of PHI-accredited Passivhaus Certifiers around the world, see www.passiv.de/en/03_certification/02_certification_buildings/03_certifiers/01_accredited/01_accredited.php

Useful tools

Bluebeam® software

Time-saving software for measuring lengths, areas and angles from pdfs.
www.bluebeam.com for information and downloads

Google SketchUp

Free basic version or 'Pro' version available to build 3D computer models of your project. Includes accurate sun-shading modelling.
www.sketchup.com for information and downloads

Meteonorm

Climate data for any site in the world in suitable format to insert into the PHPP, if regional data within the PHPP is not accurate.
www.meteonorm.com for information and download

THERM

http://windows.lbl.gov/software/therm/therm.html for free thermal bridge modelling software
www.carbonlite.org.uk/carbonlite/course_detail.php?cId=25 for CarbonLite training in the use of THERM
www.peterwarm.co.uk/resources/downloads for add-on software to THERM needed to calculate psi-values

WUFI and Delphin

Software for calculating heat and moisture transfer in building assemblies.
WUFI: www.wufi.de/index_e.html for information and download, including a free reduced version
Delphin: http://bauklimatik-dresden.de/downloads.php for information and download, with a 10-day trial option

Specialist building materials and products

Back to Earth
Natural building solutions (Devon)
01363 866999, www.backtoearth.co.uk

Ecological Building Systems
Low-energy building products and some on-site training. (Co. Meath, Ireland)
www.ecologicalbuildingsystems.com

Ecomerchant
Sustainable building materials (several branches)
01795 530130, www.ecomerchant.co.uk

Green Building Store
Green building products. (Yorkshire)
01484 461705, www.greenbuildingstore.co.uk

Mike Wye & Associates Ltd
Natural building and decorating products. (Devon)
01409 281644, www.mikewye.co.uk

Natural Building Technologies (NBT)
Natural products and information. (Buckinghamshire)
01844 338 338, www.natural-building.co.uk

Passivhaus Store
Specialist materials for ultra-low-energy builds. (Devon)
0345 257 1540, www.passivhausstore.co.uk

Tŷ Mawr
Ecological building products (Powys, Wales)
01874 611350, www.lime.org.uk

Airtightness testing

Air Tightness Testing and Measurement Association (ATTMA)
The main professional body for accrediting of airtightness testing.
020 8253 4514
www.attma.org

Air Leakage, Detailing and Awareness Services (ALDAS)
The most experienced in the UK on airtight testing.
Paul Jennings: 07866 948200, paul@gaiagroup.org

Index

Page numbers in **bold** refer to figures and tables; page numbers in *italic* refer to photographs

aesthetics 12, 27, 29, 50, 182, 183, 186
Agrément Certificates 130, 131
air changes per hour **18**, 18, 67, 105, 123, 124, 125, 126, **127**, 190, 194, 195, 196, 213
air filters 198-9, *199*, 213
air leakage 22, 123, 146, 147, 148, 189
 avoiding at construction stage 129, 133-5
 avoiding at design stage 100, 132-3, *132*
 Passivhaus standard **18**, 67, 105, 190
 zero carbon benchmark 67
 see also airtightness
air permeability 18, 44, 68, 124, 126, 131, 189, 190
air quality *see* indoor air quality (IAQ)
airtightness 22, 122-43, 189
 certification 44
 construction details 138, *138-42*
 continuity of airtightness layer 123, 128, 130, 132, 147, 156
 costs 35-6, 134
 electrical and plumbing penetrations 133, 134, 137, *142*
 EnerPHit standard 30
 and indoor air quality (IAQ) 123-5
 international standards 126-7, **127**
 junctions 130, **132**, 132, 136, *139, 140*
 materials 129-32, **130**, 146
 MVHR efficiency and 204
 Passivhaus standard **18**, 22, 123, 125, **127**, 128
 quality control unit 138
 retrofits 30, 36, 81, 131, 134, *138, 139*
 sequencing 135-6
 tapes 36, 64, **130**, 130-1, 132, 134, 135, *140-1*, 178
 tests 125, *125*, 126, 137-8, 157, 251
 weak areas 134
 windows **132**, 134, 135, *136, 138, 140-1*, 178
 zero carbon standard 126, **127**, 190
airtightness champion 134

American Society of Heating, Refrigerating and Air-conditioning Engineers (ASHRAE) 149, 192, 249
annual primary energy demand **18**, 23, 24, 44, 68
apartment blocks 24, 35, 70, 184, 192, 204, 214-15
architect 23, 29, 37
 briefing 83
 Passivhaus-qualified 43, 83
 selecting 83-4
 see also team approach
Armaflex® 106, 200
asphalt 165, 167
asthma 150, 191, 199
Austria 41-2, 47, 64, 85, 227, 228

balconies 42, *43*
bathrooms and WCs 24, 67, 104, 150, 189, 190, 196, 210-11
bedrooms 25, 54-5, 104, 190, 196
biofuels 69, 70, 72
blinds 20, 212
 external 182-3, *182, 183*
 integral 182, *183*, 220, *221*
boilers 70, 230
 combi boilers 231
 wood-fuelled 27, 70, 71, 222-3, 230-1
breathable assemblies 147, 156-7
 see also vapour-permeable materials
BREEAM EcoHomes 62, 69
brick 62, 78, **154**, 155, 157, 159, 161-2, 185, 232
building components and materials 61-5
 breathability 19, 128-9, 146, 156-8
 cyclical wetting and drying 129, 151, 160, 161, 162
 embodied energy *see* embodied energy
 hygroscopic *see* hygroscopicity
 moisture in 150-5
 natural materials 64-5, 129
 Passivhaus-certified 42, 55
 sourcing 50, 55, 61, 62, 72-3, 185, 227-8
 thermal mass 184-5
building plot 34, 75-6
Building Regulations 11, 56, 57, 65, 67, 117

Building Research Establishment (BRE) 41, 62, 109, 249
building standards, mixing 19
building-fabric-based energy standards 222-3, **223**

Camden Passivhaus *53*, 84, 234, *234*
Canada 49, **127**, 225
canopies 183
capillarity 150-1, 153, 155, 156, 161
capillary blocks 159, 161, 162
carbon dioxide emissions 65, 66, **70**, 124, 222
 carbon compliance 66, 68, 69
 Dwelling Emission Rate (DER) 57
 measuring 192, **193**
 offsetting 69-70
 regulated and unregulated 66, 68
 Target Emission rate (TER) 57
 see also zero carbon standard
carbon sequestration 65
cavity walls 78, 112, 158-9, 164, *164*, 189, 213
CE mark 95-6, 97
cellulose insulation 64, 165, 178, 232
Certification 19, 40-5, *42*, 69, 83, 133
 costs 41, 42, 86
 importance of 41-2
 of building components 42, 55
 of buildings 41-2
 post-construction 43
 process 43-4, **45**
Certified European Passive House (CEPH) course 43, 83
chi-values 115, 116
chimney flues 79-80
clay plasters 25, 151, **154**, 158, 185, 195
Code for Sustainable Homes (CSH) 11, 19, 56, 62, 65, 66, 69, 100, 223
cold bridges *see* thermal bridges
cold radiant 21, 36
comfort ventilation *see* mechanical ventilation with heat recovery (MVHR)
Community Land Trusts 47
compact units 24, 42, 105, 203, *203*
concrete 62, 78, 111, 123, **130**, 130, 157, 167, 232
condensation 111, 146, 148, 151, 162, 173, 184, 194, 195, 195-6, 200
 interstitial condensation 129, 145, 147, 155, *157*

conservation areas 76-8, 84, 165, 218
construction sector
   blame culture 133, 226
   damage–repair cycle 49, 134
   education and training 48-9, 137, 186, 206-7, 225-6
   employment structures 49, 136-7, 226
   outsourcing practice 226
   risk-averse attitudes 50
   UK building culture 47-8
contractors
   briefing 133
   Certification 43
   construction contracts 85-6
   managing airtightness issues 22, 49, 84-5, 133, 134, 137
   selecting 84-6, 133, 185-6
   team working 23
   trade sequencing 133, 135-6
   training 137
   see also team working
contracts 85-6
cooker hoods 189-90, 199
cooling load 23, 26, 97, 109
corner-cutting 54, 86, 87
cost-effectiveness 37-9, **38**
costs 12, 35, 86
   budget, apportioning 86
   build costs 23, 35-7, 54, 86, 134, 186
   capital costs 38-9
   Certification 41, 42, 86
   cost parity 34-5
   design and energy modelling 20, 86
   economics of a Passivhaus 32-9
   overruns 86
Crossway Passivhaus *3-4*, 235, *235*

Delphin 158, 251
Denby Dale Passivhaus *2*, 158-9, *159*, *182*, 213-14, 235, *235*
desiccants 151, 173
design 12, 20, 29, 51, 91
   'designing on the hoof' 86-7
   professional design team 49-50
   see also architects; Passivhaus Designers; Passivhaus Planning Package (PHPP)
deviation from north 100, **101**, 101, 108
diurnal temperature cycle 185
doors 179-80, 180, *180*
drainage plains 155, 158, 160, 164, 165
draughts 21, 36, 123, 124, 128, 189, 191, 196, 211

see also air leakage
dry air 25, 149, 194-5, 213
drying clothes 25, 195, 210
ducts 44, 54, 67, 80, 106, 199-201, **200**, 206
   insulation 200, 202
   psi values 106
dust mites 150, 195

eco-houses 56, 61, 192
economics of a Passivhaus 32-9, 71
electrical penetrations 133, 134, 137, *142*
electricity 24, 230
   grid electricity 69, **70**
   solar-generated 69-70
embodied energy 19, 27, 55, 61-4
   assessing 61-2, **63**, 63
   low-embodied-energy materials 27, 55, 61, 80, 192
energy assessment procedures 56-8
energy balance 75, **76**, 176, 181
energy costs **38**, 38-9, 209
energy in use 11, 47, 61, 62, 64, 69, 127
energy modelling *see* Passivhaus Planning Package (PHPP)
Energy Performance Certificates (EPCs) 224, *224*
energy returned on energy invested (EROEI) 71, *71*, 72
energy-poor future 11, 12, 33, 71-2
EnerPHit standard 30-1, 42, 80, 81, 82, 118, 120
entropy recovery ventilation (ERV) 195
entry and exit from a Passivhaus 211
equilibrium moisture content (EMC) 151
ethylene propylene diene monomer (EPDM) 130
European Standard (EN) 96, 192
extensions 81-2, *82*
external insulation 76, 78, 81, *115*, 128, *138*, 162, **164**, 164-5, 167, 177, 219, 221
external thermal insulation composite systems (ETICS) 164, 165
extractor fans 189-90, 192, 195, 199, 209

Fabric Energy Efficiency Standard (FEES) 11-12, 31, 58, 66-7, 69, 223, 224
fire safety 205
fires, open 25-6, 27, 189

flashing tapes 179
flat roofs 165-7, **166**
floor insulation 78-9
flow-finder device 205
foam board insulation 80
Foamglas® 35, 167
form factor **21**, 21, 80, 86, 220
fossil fuels 63, **70**, 71-2, 222
freeze–thaw cycle 161
fuel poverty 224, 225

g-values 26, 97, **99**, 176
Germany 17, 48, 85, 131, 184, 192, 204, 214-15, 221, 227
glass **130**, 150, **154**
   see also doors; windows
government policies 47, 50, 56, 61, 65, 66, 67, 137, 218, 220, 227, 228-9
grease filters 199, 206
greenwash 41
grommets, airtight *142*
ground source heat exchanger (GSHX) 184, 198
groundwater 155
Grove Cottage *53*, 214, 235, *235*
gypsum plaster **130**, 130, 150, **154**, 177, 232

health 17, **149**, 149-50
heat load **18**, 23, 24, 80
heat pump technology 203, 230, 231
heat recovery ventilation *see* mechanical ventilation with heat recovery (MVHR)
heated towel rails 24, 210-11
heating options 24, 35, 230-1
   see also space heating
hemp 25, 64, 65, 96, 232
hempcrete 157
Home Builders' Federation (HBF) 228-9
Home Energy Rating System (HERS) 41, 69
Homes and Communities Agency (HCA) 48
hot climates 17, 23, 25, 26
hot water systems 23, 24
   DHW + Distribution worksheet 107-8
   options 230-1
   pipework 23, 107-8
   solar hot water 23, 67, 70, 108
   SolarDHW worksheet 108
house plants 25, 195, 213
household sizes 34
housing stock, UK 47-8, 50, 227
humidity 26, 194, 195-6
   absolute humidity 147-8

regulation 64
see also relative humidity (RH)
hygrometers 158
hygroscopicity 25, 150, 151, 153, 155, 156, 158, 165, 195

I-beams 95, *95*, 113, 178
immersion heaters 23
indoor air quality (IAQ) 22, 25, 67, 123-5, 145, 191-6
   classification 192
   effect of RH levels on 150, 194, 195
   health impacts 64
   mechanically ventilated solutions 25, 68, 194-6
   natural solutions 22, 68
InstaBead 164
insulation 17, 20, 44, 78, 79
   blown-in 35, 96, 116, 178
   board-based 35, 78, 80, 116
   cavity-wall insulation 158-9, 189
   'closed-cell' 153, 155, 157, 167
   diffusion-resistant 200
   external 76, 78, 81, *115*, 128, *138*, 162, **164**, 164-5, 167, 177, 219, 221
   internal 76, 161-2, *162*, **163**
   natural materials 151, 153, 162
   pre-installed 35
   roll-based 116
   thermal performance 123, 151
   ultra-low-thermal-conductivity 78
intelligent breather membranes 129, **130**, 130, *142*, **154**, 160, 167
internal heat gain 23, 93, 212
interstitial condensation 129, 145, 147, 155, *157*
Ireland 41, 55, 130, 131

k-values *see* lambda values
kitchens 67, 104, 150, 189-90, 190, 196, 199, 209-10

lambda values 64, 93, 95-7, 232
land banking 34, 227
land for release 47, 48, 227
land prices 33-4
land tax 227
Leadership in Energy and Environmental Design (LEED) 41, 62, 69, 191
Leca® 156, *157*
life of a building 20, 22, 71
Lifetime Homes 62, 69
lime mortars, plasters and renders 62, 96, **130**, 158, 161, 165, 195
listed buildings 76, 77, 84, 218
living in a Passivhaus 208-15
living roofs *166*, 167

local authorities 47, 48, 56
   see also planning system
localism agenda 47, 50, 51, 218
log burners 26, 27, 70
low-carbon economy 11, 12
low-energy standards, alternative 56-8, 69

material degradation 67, 129, 134, 137, 146
mechanical ventilation systems 128, 150, 189-90, 205-6
mechanical ventilation with heat recovery (MVHR) 22, 35, 44, 58, 73, 104, 190-1, 196-207, 211
   air filters 198-9, *199*
   and air quality 25, 68, 194-6
   airflow rates 194, 195, 196, 205
   automating 192
   bypass mode 55, 184, 201-2, 204
   central unit 198, *198*, 203-4, 206
   central unit size and location 105, 106, 196-7, 200, 210
   certified models 42, 105-6, 203
   compact unit 24, 42, 105, 203, *203*
   components 198-202
   condensate drain 198, 206, 214
   control unit 202, *202*, 207, 213
   design and installation 104, 204-5, 206-7, 207
   efficiency 203-5
   ERV units 25
   fire safety 205
   frost-protection strategy 198, 199
   grease filters 199, 206
   heat recovery element 190-1, **191**
   inflexibility 54
   intake and exhaust ducts 199-200, **200**
   noise 27, **106**, 197, 201, 205
   objections to 205-6
   silencers 201, *201*, 205
   space heating 202-3
   supply and extract air terminals 201, *201*
   supply and extract ducts **200**, 200-1
   system failure 206
   transfer paths and zones 196, **197**
   user briefing 85, 207
   values **106**, 106
microporous membranes 146
mineral fibre insulation 153, **154**, 159, 232
moisture management 76, 77, 79, 129, 144-67
   in construction 155-67
   moisture modelling 81, 158

new build drying out 195
retrofits 145, 156-7, *157*, 161-5
whole-house strategy 145
see also humidity; moisture transport
moisture transport 145-55
   in materials 150-5
   liquid moisture 146, 147, 150
   water vapour 25, 76, 146, 147, 148, 150, 155, 162, 194, 195
monolithic membranes 146
mortar joints, maintenance 131, 159, 161
mould growth 67, 111, 137, 145, 146, 173, 195, 196
MVHR *see* mechanical ventilation with heat recovery (MVHR)

National House Building Council (NHBC) 48, 229
national parks 76
natural materials 55, 61, 64-5, 151, 153, 162
net energy 71, 72, 73
Netherlands 47
noise 27, **106**, 197, 201, 205, 209

off-gassing 64, 124
oil 71-2
oriented strand board (OSB) **130**, 130, 131, **154**, 156, 160, **163**, 178, 232
overheating 23, 25, 67, 69, 75, 107, 180, 181-5
   Passivhaus standard 22, 25, 181

paint products 125, 164
parging 135, *136*, *139*, 177
passive stack ventilation 67, 126, 181
Passivhaus
   challenges of meeting the standard 46-59
   concept 17, 19-23
   cost-effectiveness 37-9
   disadvantages 54-8
   economics of 32-9, 71
   life of a building 38, 71
   living in 208-15
   misconceptions about 23, 25-7, 29
   resilience 72-3
   standard **18**, 19
   UK Certified projects 234-8
   see also Certification; setting up a Passivhaus project
Passivhaus Certifiers 41, 251
Passivhaus Consultants 43, 83, 250
Passivhaus Designers 42, 43, 44, 81, 83, 250

Passivhaus Institut (PHI) 19, 49, 61, 249-50
Passivhaus Planning Package (PHPP) 20, 39, 43, 58, 69, 76, 81, 84, 86, 90-109, 176,
　accuracy of 91, **92**
　buying 92
　Final Protocol spreadsheet 104, 196
　history of 91
　software 91-2
　worksheets 92-109
Passivhaus Trust 83, 250
payback period 37-8, 39
peak net energy 71, 72
phenolic foam 64, 157, 160, 162, **163**, 165, 232
PHI *see* Passivhaus Institut (PHI)
photographic records 44, 133
planning system 50-1, 76-8, 217-21
　conservative bias 50, 76, 217, 219
　discretionary process 218, 219
　inherent problems 217-18
　inhibition of innovative design 50, 217, 219-20
　local and regional development plans 217-18, 220
　national policy 218, 220, 227
　planning advice 78
　planning criteria 219
plumbing 137
pollutants 124, 145, 191
polyethylene 130
polyisocyanurate foam 157, 232
polystyrene **154**, 157, 232
polythene 130, 151, **154**
polyurethane foam 132, 157
post-completion
　briefing the occupants 85, 207
　review 135
PHPP *see* Passivhaus Planning Package (PHPP)
Princedale Road Passivhaus 77, *77*, 161, 162, *162*, *187*, 203, *203*, 237, *237*
private finance initiative (PFI) 228
property prices 33-4, **34**
property tax 225
psi-values 80, 93, 98, 99-100, 100, 106, 116, 173, 174-5
　calculating 116-18, **117**, 120
public policy changes, proposals for 217-29

rain penetration 131, 150-1, 155, 161, 162, 164, 165, 179
rain screen 26, 95, 159-60
refurbishment *see* retrofits

relative humidity (RH) 25, 111, 129, 145, 146, 147-8, **148**, 151, 158
　and comfort levels 145, 149
　and health **149**, 149-50
　and indoor air quality 150, 194, 195
　measuring 158
　target levels 148, 150
rendered finishes 158, 160, 161, 164, 165
reservoir cladding 155, 160, 161
retrofits 29-31, 62, 78-82, 211-13, 214
　air humidity 196
　airtightness 30, 36, 81, 131, 134, *138*, *139*
　brick or stone façades 78
　cavity walls *see* cavity walls
　economic viability 56
　EnerPHit standard 30-1, 42, 80, 81, 82, 118, 120
　extensions 81-2, *82*
　external wall insulation 76, 78, 81, 162, **164**, 164-5
　fireplaces 79-80
　floor insulation 78-9
　high-embodied-energy materials 55, 63
　invasive nature of 56
　moisture management 145, 156-7, *157*, 161-5
　phasing 80-1, **163**
　planning restrictions 76
　roof insulation or repair 79, 81
　thermal bridges 20, 80, 81, 95, 111, 118-20, 162
　VAT 56, 80, 224
　ventilation systems 80
　windows 80, 81, 82, 177-8
roofs
　insulation 79, 81, *142*
　metal roofing 26
　overhangs 50, 79, *79*, 155, 158, 159, 164, 183
　roof lights 180-1
Royal Institute of British Architects (RIBA) 83, 250
rubber membranes 129, 130

schools 192, *194*
self-build sector 47, 48, 55-6, 83, 86, 217, 218, 226-7
sequencing 135-6, 137
　non-standard 132, 135-6
setting up a Passivhaus project 74-87
　architect and builder, selecting 82-6
　choosing a plot 75-6

　client role 86-7
　defining the brief 83
　planning considerations 76-8
　retrofit projects 78-82
shading 75-6, 181, 181-4, 212
　shading devices 20, 25, 26, 50, 85, 107, 181-4, 221
　Shading worksheet 101-2, **102**, **103**
　summer shading 101
　winter shading 101
sheep's wool 64, 151, 162, 232
short-termism 12, 33, 228, 229
silicate blocks 159
silicone sealants 132
sockets 133, 134
solar gain 17, 26, **76**, 97, 100, 169, 176, 189
　g-values and 97, 176
　'shoulder' months 24
　summertime 20, 101
　wintertime 20, 75, 97, 101, 176, 181
solar hot water system 23, 63, 67, 70, 108
solar panels 69, 70, 221, 222, 231
sound attenuators 27
space heating 11
　annual space heat demand **18**, 23
　EnerPHit standard 30
　options 24, 26-7, 230-1
　Passivhaus standard **18**, 58
　through MVHR system 55
　zero carbon standard 58, 66-7, 70, 127, 190
specific heat capacity 93, 107, 185
stamp duty 66
Standard Assessment Procedure (SAP) 56, 57-8, *57*, 67, 91, 108, 117, 125-6, 127
steel 61, 62, 111, 116, **130**, 130, *139*, 150, 177-8, 185, 232
stone 78, 129, 155, 161, 185, 232
straw-bale construction 61, 64, 65, 232
structured insulated panels (SIPs) 112
supply duct radiators 24, 202, *203*, 212
Sustainable Building Association (AECB) 49, 100, 118, 223, 249
Sweden 126, **127**
Switzerland **127**

Target Fabric Energy Efficiency (TFEE) 67
team approach
　briefing 136
　bringing the right team together 54
　communication 136, 137, 226
　cooperation and trust 82-3, 86

whole-team approach 23, 54, 82, 83, 133-4, 137, 138
temperature
  comfort design criteria **22**
  stability 212, 213
THERM 100, 120, 174, 175, 251
thermal bridges 17, 20, 26, 30, 35, 44, 93, 95, 110-21
  calculation 95, 111, 120
  constructional 111-12, **112**, **113**
  corner junctions 113, **114**, 115, 116
  designing out 35, 42, 80, 95, 111, 118
  geometrical 112-15, **114**, 116
  impact of 111
  linear 116
  measuring 115-18
  origins of 111
  point thermal bridges 115-16
  psi-values 80, 95, 116-18
  retrofits 20, 80, 81, 95, 111, 118-20, 162
  thermal-bridge-free 113, 116, 117
  wall–floor junctions 35, 80, 112, **113**, **115**, **117**
  wall–roof junctions 79, 80, 115, 116, 118, **119**, *120*
  windows 81, 100, 173, 174
thermal bypass 78, 112, 128, 131, 164
thermal comfort 21-2, **22**, 36, 128, 145, 149, 189, 191
thermal conductivity values *see* lambda values
thermal emissivity 169, 176
thermal envelope 18, 27, 29, 93, 111, 116
thermal mass 26, 184-5
thermal resistivity 96
thermal stores 231
thermographic imaging 17, *17*, 138
timber-frame construction 64, 65, 111, 112, *113*
  airtightness 136, 138, 157
  insulation 160
  moisture management 153, 156, 159-60, **160**
Thistle Hardwall® 177
Totnes Passivhaus *53*, 78, 118, **119**, **164**, 179, 183, 211-13, 220, 238, *238*
tradable energy quotas (TEQs) 73
treated floor area (TFA) 18, 21, 57, 93
  measurement 57, 67, 94-5, 127, 223
trees 75, 183
trickle vents 22, 67, 124, 126, 189, 192

U-values **21**, 21, 29, 93
  calculation 171-6

windows 36, 97, 98-9, **170**, 170-6, **175**, 177, 180, 182
ultra-low-energy build 29, 33, 49, 50, 51, 54, 58, 62, 64, 91, 96, 97, 98, 124, 137, 145, 148, 151, 158, 169, 174, 177, 181, 186, 189, 212, 217, 220, 222, 225
underfloor heating 211
USA 41, **127**, 151-2, 176, 191, 192, 195
  US/metric conversions 233

vacuum-insulated panels (VIPs) 78
vane anemometers 205
vapour permeability 128-9, 146, 147, 150, 151-5, 156
  1:5 ratio 156, 160, 165
  measuring 128-9, 151-3, **152**, **153**, **154**
  see also breathable assemblies
vapour pressure differential 147, 153
vapour-impermeable materials 128, 129, **130**, 151, **152**, 157-8, 160, 165
vapour-permeable materials 128, 129, 146, 147, 160, 165
  see also intelligent breather membranes
VAT 56, 80, 223-4
ventilation 148, 188-207
  comfort design criteria **22**
  and indoor air quality (IAQ) 191-6
  mixed-mode ventilation 127-8, 184, 189, 190-1
  natural 67, 68, 180-1, 190
  night ventilation (flushing) 184, 190
  noise 27, 209
  passive stack 67, 126, 181
  summer 184, 189, 190
  Ventilation worksheet **104**, 104-6, **105**
  see also ducts; mechanical ventilation with heat recovery (MVHR)
verandas 183
volatile organic compounds (VOCs) 64, 124-5, 191, 195

Warmcel 35
water usage criteria 68
water vapour 25, 76, 146, 147, 148, 150, 155, 162, 194, 195
weep holes 158, 162, **163**
wind-tightness 128, 131, 147
wind-washing 128
windows 168-87
  aesthetics 182, 186

airtightness **132**, 134, 135, **136**, *138*, *140-1*, 178
argon-filled units 99, 169, 173, 176
Certification 42
conservation-friendly 77
costs 36-7, 186
curved **101**, 101
deviation from north 100, **101**, 101
doors 179-80
double-glazed 36, 170
embodied energy 63-4, *180*
fixed 36-7, 172
frames 98-9, 100, 177, *177*
g-values 26, 97, **99**, 176
installation 81, 99-100, 134, 135, 173-8, *178*, *179*
inward opening 169-70, *170*, 178
krypton-filled units 99, 169
low-emissivity coating 169, 176
mullions 100-1, 102, 221
opening 22, 25, 54, 55, 180-1, 189, 190, 192
ordering 185
psi-values 98, 99-100, 118, 173, 174-5, 181
retrofits 80, 81, 82, 177-8
reveals 135, 177, 178, 181
roof lights 180-1
sills 179, *179*, 180, *180*
solar gain *see* solar gain
spacers 36, 151, 169, 173, *173*
thermal bridging 81, 100, 173, 174
tilt-and-turn 169-70, *170*, 184
triple-glazed 36, 64, 98, 135, 169, 170, *172*, 185, *187*
U-values 36, 97, 98-9, 100, **170**, 170-6, **175**, 177, 180, 182
window schedule 100
xenon-filled units 99, 169
wood burners 26, 35, 230, 231
wood fuel 24, 27, **70**, 70, 71, 222
woodfibre board 64, 129, **154**, 156, 162, 165, 178, 232
WUFI 158, 251

Zero Carbon Hub 58, 66, 250
zero-carbon standard 11, 27, 31, 51, 61, 65-71, 126, 206-7, 222
  air-leakage benchmark 67
  airtightness target 126, **127**, 190
  Passivhaus–zero carbon relation 65-71, 223
  space heating target 127
  and stamp duty 66
  sustainability criteria 67-9
  zero carbon definition 65-6